S0-FLS-048

DATE DUE

OCT 08 1985		
NOV 07 1985	APR 23 1986	
FEB 18 1986		
MAR 14 1986		
JUN 2 1986		
AUG 25 1989		
201-6503		Printed in USA

AUG 26 1985

INVESTING IN OIL

INVESTING IN
OIL

by

Walter Cook

ROUNDTABLE PUBLISHING, INC.
SANTA MONICA CALIFORNIA

Copyright © *1985* by Walter Cook

All rights reserved. No part of this book may be reproduced in any form or by any electronic or mechanical means, including information storage or retrieval systems, without permission in writing from the Publisher,

ROUNDTABLE PUBLISHING, INC.
933 Pico Boulevard
Santa Monica, CA 90405

Photographs and drawings appearing on pages 18, 20, 21, 48, 50, 52, 71, 73, 78, 107, 115, and 130 are reprinted with permission from *Lessons in Well Servicing and Workover*, copyright 1972, the Petroleum Extension Service (PETEX), the University of Texas at Austin.

First Printing, *1985*

Library of Congress Catalog Card Number — 83-63203

PRINTED IN THE UNITED STATES OF AMERICA

This book is dedicated to all of the good, honest people in the oil and gas business, with a remembrance; drilling and completing a good gas or oil well is not like going into a store and buying a can of beans.

CONTENTS

Part I — LOOK BEFORE YOU LEAP

Chapter 1
The Mineral Lease, 3

Obtaining a Clear Lease • Mineral Rights without Surface Rights • The Resident Lessor • Hidden Costs in Mineral Leases • Overriding Royalty Interests • Working Interest • Percentages and Disbursements • How Investments Go Sour • The Fast-Buck Artist • Determining Good and Bad Leases • Investing Time before Investing Money

Chapter 2
Small Companies and Private Corporations, 25

Behind the Public Relations • Small Companies on the Stock Exchange • Employee Stock Options • Abuses of Stockholders' Money • Private Corporations

Part II — DRILLING PROSPECTS

Chapter 3
Working Interest in a Drilling Venture, 43

Giving Away More Than You've Got • Improving Someone Else's Property • Reserves • Depletion

Chapter 4
Re-Entering Old Holes and Wells, 53

Supply and Demand • Old Well Logs • New Technology • Plugging Holes vs. Plugging Wells • The Dangers of Re-Entry

Contents

Chapter 5
Producing Properties, *67*
    A Common Scam • Check the Run Tickets • Workover • Liens on Producing Properties • Get All the Information • Perforating Procedures

    PART III — PROTECT YOUR INVESTMENT

Chapter 6
Company Expenditures and Your Money, *83*
    The A.F.E. • Companies Furnishing Their Own Rigs • Overcharging for Pipe • Coring • Consultation • Gas Pipelines • Hauling Oil

Chapter 7
Incompetence, *95*
    Contractors' Errors • Easements • Follow-Through • Incompetent Consultants • Investigate for Competence

Chapter 8
Defining Responsibilities, *111*
    Irresponsible Cuts in Equipment • Blowouts • Used Equipment • Quick Decisions • The Insurance Representative

Chapter 9
Investing Points and Questions, *123*
    Tax Incentives • When Do You Turn Over the Money? • How Far Will Your Money Go? • Are You Paying Someone Else's Way? • When Is a Dry Hole Not a Dry Hole? • Are There Any Loopholes? • Further Questions to Ask • Finding the Good Guys

Glossary, *137*

Index, *181*

INTRODUCTION

This book has been prepared with one objective. To inform potential investors of the processes of oil and gas exploration, of drilling procedures, and the resulting successes of production and sales of oil and gas. Of necessity, this involves examination of many failures of such procedures, of the losses of investment through the pitfalls of drilling and production, and the many lessons of incompetence that attend oil and gas ventures.

It is my aim to produce within potential investors an awareness of the whole operation that makes up an investment in gas and oil ventures, whether it is one well or an entire field of wells that will, in the end, produce the product and be marketed so that the investment receives a return and ultimately a profit.

Many investors are aware of the great amounts of money that can be realized in the oil and gas business. Some new investors believe that all it takes to make a good gas or oil well is a drilling rig, a piece of land, a crew to operate the equipment, and a few knowledgeable souls who can form a company and start digging.

It would make oil exploration simpler if we could take a feathered dart, throw it at a map on the wall, and start drilling a proposed well on the site where the dart fell. Imagine someone shouting: "Get a drilling rig and start there!"

I look back over more than thirty years of jobs and associations with many oil and gas companies. It is impossible for me to count the number of drilling deals I've been involved with or heard about; some were just too good to be true, while others weren't worth one minute of some promoter's hoopla. I have attended many schools, studying

Introduction

oil and gas related subjects and operations, and I have participated in scores of business seminars. All of this education has been combined with a thorough working knowledge and experience through these many years. I feel my accumulation of experience and knowledge might be helpful now to give some insight to investors, to banks and attorneys, even to mineral rights owners and landowners.

I am asked frequently: "How can I get into the oil and gas business and make a lot of money?" Of course, anyone can play the investment game, but it's better to play it for profit rather than loss. I am hopeful that in the chapters that comprise this book there are some guidelines and pertinent information for the experienced investor as well as for the potential investor.

Profits beyond our imagination can be gathered in the oil and gas business — by the same token, just as much loss can be weathered. There are as many variables in investment successes as there are in failures. The road to high profit is hazardous, but with proper investigation, with doing the "homework," and with experienced suggestions, hazards can be minimized.

In every trade and profession there is a special language. In the oil and gas business the technical language and the slang terminology used is part of our consideration. At the end of the book we have placed a glossary, and this should prove of value to those embarking on a drilling venture or upon investments in oil and gas. To aid the reader we have often placed an asterisk (*) at a word describing an operation in exploration, drilling, production, etc. It is to be noted that various calibers of people are attracted to the rich stakes of the oil business. It is my experience that the fast-talking promoters (or "marketing agents," as they prefer to be called) are not always knowledgeable in the language of equipment and of the field-workers. Such lack can be of advantage to the potential investor.

The con artist promoter will seem to have some working knowledge of some things such as tax incentives and details concerning tax write-offs, etc., because they are aware that this is the one question about which most

Introduction

investors will inquire. If, having studied the glossary, the serious investor is able to spot the fast-talking promoter before making a firm investment commitment, then our aim to educate the investors has been reached.

I do not wish to discourage the potential investor from any investment consideration. There are more honest people in the oil and gas business than dishonest people. Although oil companies are in competition with one another, they never fail to aid fellow investors in trouble, even if it's only supplying information regarding an investment hoax. Dedication is the common bond found in people who work in oil and gas. Some are required to be away from their homes for lengthy periods, working seven days a week. Too, cold weather and extreme heat-filled days are factors that simply come with the territory. The working conditions alone are enough to discourage any half-hearted person from becoming a roughneck. But it is a challenge to many to find and pull oil and gas from the ground and sell it to the consumer.

A drilling rig set up on a location to drill down 5,000 feet is, in reality, one mile away from the work force. At 10,000 feet, the work force is nearly two miles away from their production. Those who work two, three, or five miles away from their job-site must have expert knowledge of what they are doing because they cannot make a visual contact with the goings-on deep in the well hole. If you assume you can call a department store on the telephone and ask the sales clerk to describe the man's suit on the rack, you will have to imagine — to picture in the mind — how the suit actually looks by the experienced verbal description given by the clerk.

By the same token, drilling and completion of good wells takes skilled and imaginative people who have qualified themselves with right drilling techniques and qualified crews to render a safe and economic result in producing their product. If we consider the depths of the hole, there are countless items that can go awry: the crew can drill a crooked hole and miss the oil zone, or twist off the drill pipe, or lose mud circulation, or experience a blowout, or stick the

Introduction

drill pipe. But correcting the problem is the job of the experts and crews. Facing the problems so that the oil will come up and into production tanks and into the sales lines means a completed well — and almost always, money which, in time, is profit (after expenses and taxes are met).

It is incumbent on the investors to choose the right people to listen to and the right companies to place their investment money into; knowledge of the oil and gas operation must be attained so that the answers to the questions will not fall into a vacuum, the mind of the questioner.

Although we discuss here many losers and losses, I am pleased to report there are more winners (and millionaires) from aware investments. Our premise here is that to be an aware investor is to beware.

INVESTING IN OIL

Part I

LOOK BEFORE YOU LEAP

Chapter 1
THE MINERAL LEASE

You are interested in the lucrative oil and gas business and you want to be involved as an investor. The first step toward your goal is to obtain a tract of land to drill upon, or at least explore for oil and gas. The right to explore for oil and gas on a defined piece of property is, in 90 percent of the cases, purchased. One purchases rights to explore for oil and gas from a lessor, and in most lease agreements, a per-acre bonus is paid to obtain the lease. A rental fee is paid on a yearly basis to hold the lease until drilling operations begin, or the lease agreement expires. Terms should be well defined in the mineral lease agreement and understood by the persons involved; i.e., the "seller" (lessor*) and buyer (lessee*).

Obtaining a Clear Lease

For about 10 percent of the cases, the right to drill or explore for oil and gas is leased on a drilling contract basis, with no money changing hands between the lessee and the lessor. This lease agreement usually consists of the lessee's right to drill or explore for oil and gas on a piece of land within a specified amount of time. Most of these agreements are for brief periods of time (six months) and are usually made with just one lessor who owns the surface land and the mineral rights. However, the drilling contract method of acquiring a mineral lease is becoming more rare each day and more than likely will become obsolete in the near future.

LOOK BEFORE YOU LEAP

With either of the two methods of acquiring a mineral lease, 100 percent of the leaseholders must be signed into the lease agreement. If this is not done and a drilling operation is initiated, it will be considered illegal unless provisions have been made for an exception. For example, there may exist a tract of land with 25 legal heirs owning the oil, gas, and mineral rights. A lease agreement is signed and recorded with 23 heirs' signatures. Possibly two of the heirs have been overlooked or omitted by error. One might believe he has obtained a clear mineral lease. Only a 23/25 lease has been obtained. This amounts to 92 percent of the rights to explore for oil and gas. A great deal of leg-work and record-checking is necessary to obtain a clear lease. I am familiar with one lease agreement that required 364 signatures in order to get 100 percent of the mineral lease.

Mineral Rights Without Surface Rights

One of our considerations regarding an unfavorable mineral lease arrangement is to examine the aspect of a lessor who owns the mineral rights but does not own the surface land. The party owning the surface land (but not the mineral rights) will not receive any revenue from a well producing oil from beneath his land. This is considered an unhealthy working relationship between the landowner and the oil company. If a clear mineral lease is obtained in this type of situation and a drilling operation gets underway — look out! Wherever the location for that proposed well is staked, it becomes the most valuable piece of land in the world, at least in the view of the landowner. Then the road to the proposed well's location suddenly appears to have been at some time a site of great historic import — perhap's "a landing strip for flying buffalo." Naturally, the price of the land will suddenly take an upward jump. Once a right-of-way is obtained and a well is drilled and completed, with a christmas tree* installed, the landowner may complain that the rig was responsible for scaring a cow to death (or else, she walked into the darn thing and suffered brain damage). This might have been the only cow of its kind in the world — and that increased the landholder's damages even further. A year on the mineral lease passes. The landowner approaches the oil company with a complaint: "You know, that road that goes to your oil tank battery sure is dusty. That dust is blowing on my grass, and my cows are poorly. I think they're getting sand belly."

Remember, the oil company has been there for one year; the cows were poor when the oil company arrived on the scene, and they still suffer from lack of food. (The landowner has been shaking an empty feed sack at the cattle every now and then so that the oil company will think they have been fed.) But it boils down to the same old story: pay something or go to court. Anyone can see how endless the process may be. This "payment," or legal suit will be added costs, and cause adverse working conditions.

The Resident Lessor

Another situation that can be involved, and creates unnecessary problems on a mineral lease, arises when the lessor owns the mineral rights, and lives on the land he owns. Let's say this particular lessor has never even seen a drilling rig — he's raised sweet potatoes for 40 years — but now, an oil company moves a drilling rig on his property. If the company makes a dry hole, the lessor will, in all probability say, "I knew you was a-drilling in the wrong place, shoulda drilled over there." And he points off across his fields. However, if the company completes a good producing well, the landowner will, in all likelihood, become a 60-day expert. He could resort to telling the oilfield workers caring for the well what they should do. Not only that, he may go into town to tell every person he meets, "The oil company is messing up my well; I can get more oil outta that thing than they can."

LOOK BEFORE YOU LEAP

Now, if this type of lessor cannot get satisfaction from the field personnel, he will call the president of the oil company. When that happens, imagine the repercussions.

Hidden Costs in Mineral Leases

As I stated previously, most mineral leases are acquired by the rental-plus-bonus method. The bonus money paid per acre may run from $25 to as high as $1,500 an acre. The bonus amount will depend on how much oil and gas activity surrounds the lease land-area at the time the lease is obtained.

We have discussed the methods by which a mineral lease may be obtained and some of the situations that may occur affecting the cost of operations on the lease. Now we take one further step to explain some of the money transactions that may occur.

Let's say a promoter puts together a 500-acre mineral lease at the price of $50 per acre. To obtain the lease his total investment amounts to $25,000. He turns and sells the lease to an independent oil company for $100 an acre, and he keeps a 2 percent overriding royalty interest for himself. (Later, I will explain *overriding royalty interest* when we define all interest in a well.)

So, the 500-acre mineral lease has a $50,000 price tag attached. The oil company now tells its marketing people to begin selling *working interest* in the lease. The cost will be $150 per acre. This makes the original $25,000 lease $75,000; also, the oil company will keep some overriding royalty interest for themselves, if at all possible. Knock off $10,000 of the $50,000 for legal fees and marketing costs. That leaves $40,000 profit a potential investor will pay for, if buying into this drilling venture.

A potential investor must ask questions. Where did the lease come from? How much was the original cost? One must request a review of the original leases or their certified copies. Obtain the name of the lessor so that he may be called if some questions go unanswered. No one is required to pay for any excessive profits on a mineral lease.

The Mineral Lease

In this lease and one well, let's say you have decided to invest $10,000. The oil company is selling the working interest in the well for $10,000 for a 1 percent interest, paid through the completion stage. Assume you and other investors did not check out the lease costs or obtain any pertinent information, but went ahead and blindly put the monies asked for into the well. Then all of the investors paid an excess profit of $40,000. This would amount to 4 percent working interest in the well that you will never see again.

This is the way the process was worked on the investors. The promoter brought the drilling venture to the oil company. The exploration people within the company liked the prospect and thought it would sell. The promoter told them he invested $25,000 for the lease purchase plus $5,000 in legal fees. He is asking for $50,000 plus 2 percent overriding royalty interest.

The oil company agreed to take the lease under the following conditions: the promoter agrees to $30,000 cash up front plus the 2 percent overriding royalty interest. Then when the entire deal is sold, he will be assigned a 2 percent working interest in the well, free and clear. Of course, the promoter will take it because his expenses are paid for on the lease, and he is to receive 4 percent interest in a potential well. On the other hand, the oil company requires the promoter to invest his profit in the well, and the investor will, in essence, pay for it. They add $25,000 to the lease, minus $5,000 for marketing, and you, the investor, have helped pay for another 2 percent working interest, which will go back to them, the oil company.

Overriding Royalty Interests

Now I will explain the way a mineral lease or well is most often divided among different interest groups for payment from the sale of the oil or gas taken, and who carries the cost burden. This will inform an investor as to what interest of the revenue he pays for and how all the revenue is disbursed.

Let's continue with the original mineral lease that contained 500 acres. You have finally obtained a 100

LOOK BEFORE YOU LEAP

percent mineral lease, signed and recorded, for the right to explore for oil and gas. In almost all of the cases I am familiar with, the person from whom the lease was secured was able to keep 1/8th or 1/6th of the original 100 percent. This is referred to as the *lessors' royalty*.

Next, in the division of the interests from the mineral lease, there is the overriding royalty interest. This is interest payments on the revenues from the sale of the oil and gas with no working capital involved. It is usually a bonus, a favor for (or through) friendship. Remember the promoter wanted 2 percent overriding royalty interest as a bonus; and he received it. Also, the leasehound (land person who obtains the lease) will receive a small interest as a bonus. A banker may share in the overriding royalty interest for steering the marketing agent to persons with money to invest. Most of the time, the oil company will give persons in top management this same interest bonus, or else the company will have a royalty pool. At one time, I was one of 10 people working with an oil firm involved in such a royalty pool.

The oil firm held onto as much overriding royalty interest as they believed the lease would stand, before placing it on sale to investors. On our original 500-acre lease, the company placed 3 percent interest in a royalty pool. Since there were 10 of us in the royalty pool, each of us had free and clear 1/10th of 3 percent. This may not seem like very much to an outside investor, but, in one year, I obtained an interest in 21 separate leases as a bonus. This kind of income will last as long as there is a producing well on the leases. It can be sold or money can be borrowed against it. But, again, as the investor, you are stuck for paying all overriding royalty interest revenues.

Working Interest

The last "interest" of the 100 percent on the 500-acre lease is the *working interest*. This interest (in a well) is the portion in which one is investing. If you are the investor,

The Mineral Lease

you are actually carrying the lessor and the overriding royalty interest. (I will explain later what each investment point means. However, at this time, you need to know the boundaries you are working within.) Read the next passage carefully; it is important!

A promoter could approach an investor and use different sales angles:

    a) Do you, as an investor, want an interest to the logging point or the casing point?

    b) Do you want an interest to the completion point?

    c) Do you want an interest through the completion point plus oil into the tanks, or at a sales point for the gas?

Most important, the promoter will attempt to sell you, the investor, an interest in *one* well on the 500-acre lease, not the entire lease. Promoters have an established reason for selling one well at a time.

Another selling point parlayed to the investor is *reserves*. The promoter will confirm that the oil and gas reserves are there. Using a very sound approach, a well is not a good well until oil is going into the tanks or gas is moving through the sales line in an economical manner.

Percentages and Disbursements

Assuming the 500 acres is leased and a free and clear title to a 100 percent lease has been obtained, let's examine the money moving into the hands of the investor, the lessor, and the *overriding royalty interest* owners.

ORIGINAL LEASE		100 %
Lessor (good lease) 1/8	12.5%	
Amount left before overriding royalty		87.5%
Overriding royalty interest to:		
Promoter	2 %	
Oil company royalty pool	3 %	

LOOK BEFORE YOU LEAP

Leasehound	1 %	
Chairman of Board, oil company	1 %	
Marketing agent	1 %	
Engineer furnishing geology to promoter	3 %	
Attorney for legal work	2 %	
V.P. of exploration, oil company	1 %	
Financial advisor to investor	2 %	
Total overriding royalty interest	16 %	
Working interest remaining		71.5%

This 71.5 percent *working interest* now is stated as "100 percent working interest." Simply, this means that 100 percent of working interest in the well will in fact receive only 71.5 percent of the sales revenue, minus taxes that must be paid. The promoter will sell the investor 1 percent working interest for $10,000. By selling the working interest this way, it will create one million dollars ($1,000,000) for drilling and completing a 10,000-foot well. As an investor, you purchased 1 percent in this well and would receive .00715 percent of the revenues (minus taxes and operating expenses).

Let's suppose an oil well has been completed on this lease and it's producing 100 barrels of oil per day that sell at $30 a barrel. Total revenue each day is $3,000 gross. Listed below is the revenue disbursement for one day:

Lessor	12.5%	$	375.00
Promoter	2 %		60.00
Oil company royalty pool	3 %		90.00
Leasehound	1 %		30.00
Chairman of Board, oil company	1 %		30.00
Marketing agent	1 %		30.00

The Mineral Lease

Engineer employed by promoter	3 %	90.00
Attorney, for legal work	2 %	60.00
V.P. (of exploration), oil company	1 %	30.00
Financial advisor to investor	2 %	60.00
Total free and clear, paid by investor	28.5%	$ 855.00
Total payable to investor(s) (excluding taxes or operating expenses)	71.5%	$ 2,145.00
Total revenue per day (excluding taxes or operating expenses)	100 %	$ 3,000.00
One month's (30 days') revenues		$90,000.00

How Investments Go Sour

Now, let's review this drilling venture and visualize where the investment might have gone wrong. Actually, not wrong, just sour. This is what the right kind of investigation into the drilling venture could have saved an investor; by the same token giving the investor more returns on the amount of money invested.

First, the mineral lease price was escalated and the investor helped pay for 4 percent working interest to someone else. Income from the 4 percent interest will never be realized by the investor.

Second, in the example, I used a good lease agreement with the lessor, which was a *1/8th agreement*, or 12.5 percent of the revenue obtained from the sale of the oil going free to the lessor. If the lessor had been in a hot area, he would more than likely receive 1/6th (16.667 percent) of the revenues plus a small overriding interest as a bonus. That is, if he was leasewise.

LOOK BEFORE YOU LEAP

Third, and most important, the promoter may have led the investor to believe he would receive 1 percent of the total 100 percent revenue ($3,000 a day). The promoter sold to the investor 71.5 percent of the revenues (minus taxes and operating expenses). If an investor had purchased 1 percent *working interest* in this drilling venture and the well produced 100 barrels of oil each day that sold for $30 a barrel, the investor's daily revenue would be $21.45 (minus taxes and operating expenses).

Now then, look at 1 percent *overriding royalty interest.* The daily revenue is $30 minus taxes, free and clear of all operating expenses. In a 30 day period, the investor has taken in $643.50 but he still must pay his prorated share of the operating expenses, minus taxes. The overriding royalty interest of 1 percent has amounted to $900 in a 30-day period. This leaves 1 percent overriding royalty interest, making $265.50 more in a 30-day period than the 1 percent *working interest* for which the investor paid $10,000, and still he pays his share of the operating expenses.

The Fast-Buck Artist

Now, we return to the same oil and gas mineral lease to show how another deal can be worked. Most oil companies have a land person and a vice president of exploration. The land person's responsibility is to look over the lease with regard to the legal aspect — he is more familiar with the different land areas for profitable drilling prospects. The vice president of exploration is one who usually confirms or rejects the land deal.

Let's say that a promoter is a friend of a land person at an oil company and brings him a drilling venture to examine. It's a 300-acre mineral lease that actually does not look very good. The prospect of making a well seems slim, due to the fact there are no producing wells in the vicinity of this piece of property. There are some wells on all sides of the property but a great distance away.

The Mineral Lease

But something has attracted the land person's attention. The lease is cheap at $20 an acre, and it's a 1/8th lease from the lessor and with an overriding royalty interest of only 3 percent. Subtract the 12.5 percent to the lessor and 3 percent overriding royalty interest, this leaves a lease with 84.5 percent interest remaining. Now, bear in mind, the land person and vice president of exploration are good friends. The land person confirms with the promoter that this lease isn't too good. "But if we work together, we can make money out of this thing." And the promoter wanting a fast buck will sell the lease and agrees to help any way he can. (He is one of those "good ole boys.")

The land person takes the lease to the company's vice president of exploration and explains that, although the lease isn't that promising, they will be able to make money. It's a cheap lease with 84.5 percent interest left to work with. They can suggest that the promoter jack the price up to $45 an acre (which is still reasonable) and give each of them 2 percent overriding royalty interest. The land person will get a $10 an-acre-kickback. This still leaves 80.5 percent interest. The vice president of exploration agrees to the deal.

The land person returns to the promoter and tells him to have two *overriding royalty interest assignments* drawn up and recorded at the local courthouse. Two percent will be assigned to the land person and 2 percent will be assigned to the vice president. The promoter will have two checks drawn: one for $3,000 to the land person, and another for the same amount to the vice president of exploration. In this transaction, the promoter made $5 an acre profit, which amounts to $1,500. And he turned the lease. Then, he calls the land person with everything in order, ready to close the deal. They meet, and the promoter gives the land person all the paperwork to look over, along with two envelopes containing the two checks.

The mineral lease agreement now states that it is cut down to 80.5 percent interest. It also states there is a 7 percent overriding royalty interest already assigned and recorded. One-and-a half percent goes to the promoter and

—13—

LOOK BEFORE YOU LEAP

1½ percent goes to the engineer who helped work up the lease.

Now, I'll give an example of some fancy paperwork. There is an affidavit with this lease stating that a 4 percent overriding royalty interest has been duly assigned and recorded. The original affadavits have not been mailed back from the courthouse at that time but will be sent to the oil company as soon as the promoter receives them. These two overriding royalty assignments will be put in the lease records by the land person after the other interested parties examine the lease. However, if a well is made on this mineral lease, these records will show up, because division orders have to be made for money disbursements. The land department handles the division orders. The land person now gives the promoter a check for $13,500 for the lease. The price of the lease has jumped $7,500, and the 4 percent overriding royalty interest has disappeared. Not only have the land person and the vice president cheated the investors and the company, they have also cheated their fellow employees. This company has a royalty pool, and since both of these men are in it, they have another interest in the lease.

If you recall, this lease did not look very good from the start, but it has to be drilled. How? There must be an attractive selling point, or at least a facsimile of one. So, the exploration people and the promoter have come up with an idea that always baits the big spenders — cheap investment, big return. Remember there is 80.5 percent interest left in this lease.

The promoter takes 1 percent overriding royalty interest, and 2 percent overriding royalty interest is put into the company pool. This leaves a 77.5 percent working interest. (You will recall, the first drilling venture discussed had only 71.5 percent working interest. The 77.5 percent working interest is larger by 6 percent, and this certainly looks enticing to the investor.)

Now, one of the engineers in the exploration department has a friend who is in the contract drilling business. He asks his friend if he wants to drill this well on the 300 acres;

The Mineral Lease

he can also receive interest in the well. The contractor goes to the oil company office to look over the venture. He takes one glance and states he wouldn't put his money in it. The engineer asks his friend how much he receives a day for a rig large enough to drill this well. The contractor tells him the price is $7,200 per day, plus fuel.

How many days will it take to drill the well? The contractor estimates 20 to 21 days. The engineer gives his okay and proceeds to set up his A.F.E. (Authorization for Expenditures) in the following way. He lists 21 drilling days at $8,200 per day plus fuel; and this, of course, will be charged to the investor. The contractor is then asked to invest the $21,000 overpayment he is to receive back into *working interest* in the well. He cannot lose because he's keeping his rig busy (for $7,200 a day) and he is getting some free interest in a well — if one is completed.

Now, this contractor has a friend who is an independent trucker. The contractor tells the engineer to call his friend, the trucker. They might work out a deal whereby the trucker will get an interest. Here's another *good ole boy* for you.

The engineer calls the trucker, suggesting he move the rig to the location. The trucker agrees, and a contract is signed; the deal is sealed. The engineer asks the trucker to put an extra load on his invoice, and the oil company will pay him for it. However, the oil company wants the trucker to invest his overpayment in this oil well.

The trucker accepts. He can't lose either.

At the end of that conversation, the trucker suggests the engineer contact a friend of his who is a dirt contractor. The dirt contractor's responsibility includes building the lease roads, the location, and the pad for the drilling rig. He will also dig the mud and water pits. The contact is made, the dirt contractor agreeing to the project (Also he is willing to raise his price up $10,000 higher and to invest the overpayment in the well.) The engineer tells the promoter the profitable news and adds that it will be three or four days before the written contracts are ready. On the fifth day, the engineer should have the A.F.E.* prepared for the promoter. (An A.F.E. gives a description and cost of each

LOOK BEFORE YOU LEAP

item in this drilling venture, the determination of the cost of each 1 percent working interest to the investors.) The promoter is now ready to go to work with the "right kind of information" to present to the investors.

Would you as an investor think the scheme will sell even though the drilling prospect doesn't look profitable? Sure it will. And here's how it's done.

The promoter calls all his prospective clients, now that things are lined out, and he proceeds with a convincing sales pitch. Let's listen in on a typical conversation between the promoter and an aspiring investor.

"Hello, Mr. Jones (investor). How are you today? Listen, I have a drilling deal coming through in a few days, and in my opinion, it doesn't look like much of a deal. There are 300 acres in the lease, with no oil or gas production nearby. Funny thing, though. The drilling contractor, trucking contractor, and dirt contractor are all taking some working interest in this deal. Now they have all worked in this area before, so I believe they must know something that I don't. All three of these men have been around many years, and they should know the oil and gas business. In fact, I think I'll buy a small amount of interest in the well. Oh, I forgot to tell you there is 77.5 percent working interest. No, Mr. Jones, I don't know what 1 percent will sell for yet, but I'll find out and have it ready to sell in a few days. Call me back in three days . . . ."

The promoter places call after call to prospective investors, using this same pitch. I'm sure by now anyone can see what is happening. The people putting this operation together have to make this 300-acre lease appear good enough to sell. At first glance, they have a small selling point, which is the 77.5 percent working interest. The coup de grace is the fact that the three major contractors are taking working interest in the well. The fact that these men have been "in the business" for a lengthy time and know the area does not hurt the cause either.

The investors will never know the contractors were getting a free ride from the overpayments. The promoter conveniently neglected to mention the relatively inexpensive price of the 300-acre mineral lease. You may use this as a guideline most of the time. If a lease is cheaply priced, either the lessor doesn't know what the price should be or the leasehounds have not been working the area. I might add that when the lease men are working an area heavily, or when news of a good producing well has leaked out, the cost per acre in that area automatically soars.

Later, I will develop another angle that can be schemed with the drilling contractor. To be more explicit, it's termed "kickback."

Determining Good and Bad Leases

Having discussed a lease on which the prospects of completing a well did not appear profitable, we can look at some of the items that determine a good or bad mineral lease. It is a fact that you cannot drill and complete a well that will produce in an economical manner on every square foot of this earth. The geological composition of the different earth strata or formations is a fascinating, complex, and always changing study. Most dedicated geologists will look at and study any unusual piece of rock or earth they can get their hands on. In attempting a decent explanation of how to find a profitable mineral lease, I'll use a layman's explanation of the little geology I profess to know.

Certain strata or formations are known to be carrying oil or gas, or possibly both. These formations are most commonly referred to in oilfield terms as production zones. These zones have their own names and are too numerous to mention. A complete list would take several pages, but among the most familiar names are the Austin Chalk, Wilcox, Edwards, Cotton Valley, Ellenberger, Big Saline, and Frio. Also, there are Marble Falls, Travis Peak, James Lime, Barnett Shale, and Caddo, just to name a few.

That certain strata or formations are known to conceal oil and gas (or both) does not, however, mean they are productive in all areas that are encountered during drilling.

LOOK BEFORE YOU LEAP

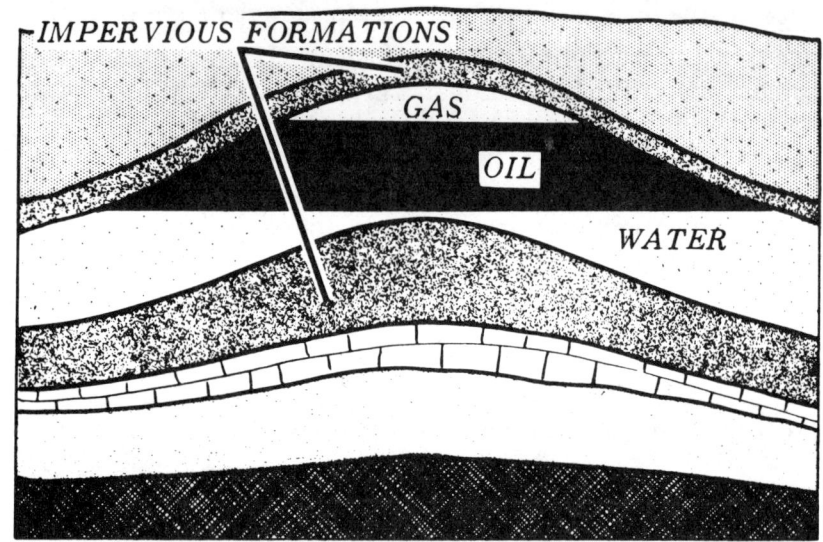

Dome type oil-bearing structure. All formations are structured with gas as top layer below earth surface, oil in middle layer, water at bottom.

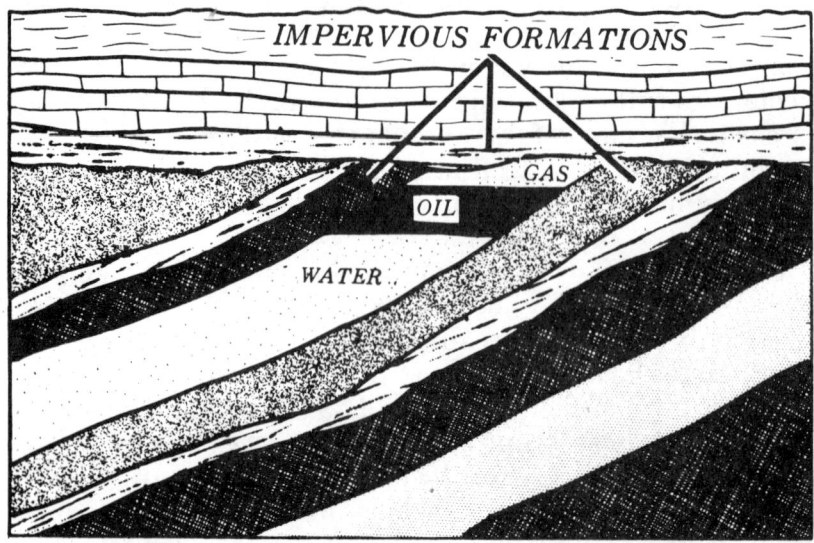

Stratigraphic type trap showing trap formed by pinch-out of the oil sand against an overlying impervious formation through which the gas, oil, and water could not pass.

The Mineral Lease

This is due largely to the formation of the earth's structure. Any formation that carries gas, oil, and water will be structured in the following way. The gas is on the top layer, oil is in the middle layer, and water is at the bottom. Picture, if you will, a glass jar, put in one-third water and one-third oil, and the last third is air. In the ground, the one-third air is gas. If gas, but no oil, is encountered, the gas is on top and the water is still on bottom. If oil, but no gas, is encountered, the oil is on top and the water is still on the bottom. In its natural state, all natural gas carries with it water. The gas must be run through a dehydrator to lower the water content for commercial use. Gas cannot carry more than seven pounds of water per million cubic feet. In fact, most good oil wells will bring some water with the oil. Water is forcing itself upward as the weight of the oil is taken off or the volume displaced.

Faulting takes place, or has already taken place, in the structuring of the earth. A fault in the earth's structure is a shifting of the earth in certain places. To illustrate a fault, hold two 3-foot pieces of board in your hands. Place one board in each hand and then place them on a table so that they will be standing upward. Touch the two boards together and keep them level at the top. Move the board in the left hand up about one foot and hold the board in the right hand still. You have just created a fault; the shifting of the boards equates the shifting of two parts of the earth.

Faults in the earth's structure cannot be seen from the earth's surface. They are usually detected by seismic survey, or well logs. A well log is a graph or sketch that is drawn showing geologists and engineers what they have come up against in the hole they are drilling. This graph or sketch is produced by running an electronic tool down the open hole; the tool sends back pulses to ink pens that draw a picture of what is down the hole. This is a fascinating science because this picture will illustrate the thickness of each zone, the porosity, the permeability, and the structure of the entire formation. The procedure for locating a fault with well logs is to employ the use of several logs in a concentrated area.

LOOK BEFORE YOU LEAP

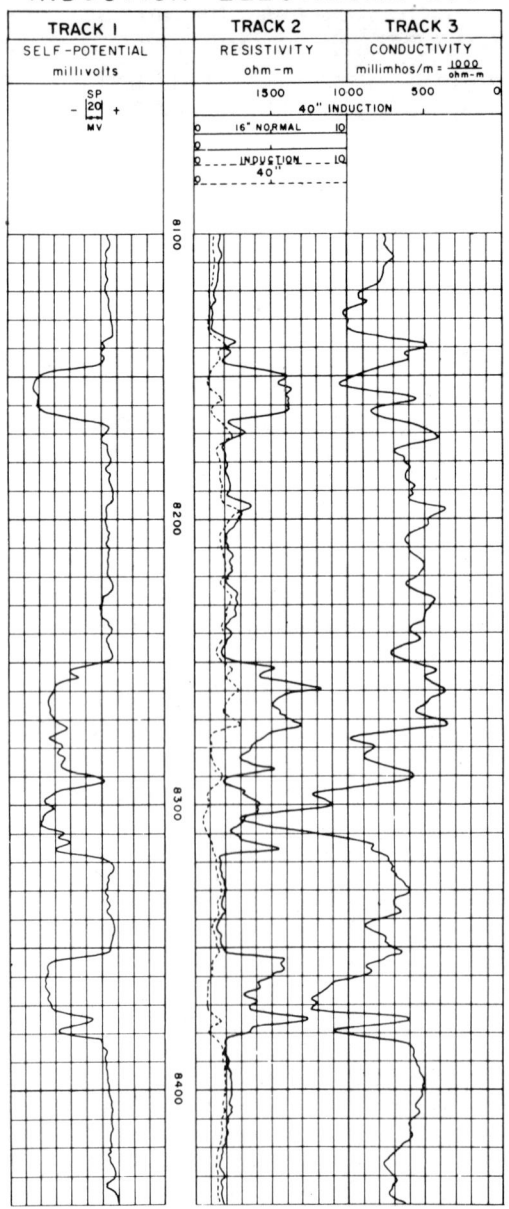

General appearance of an electric log.

—20—

The Mineral Lease

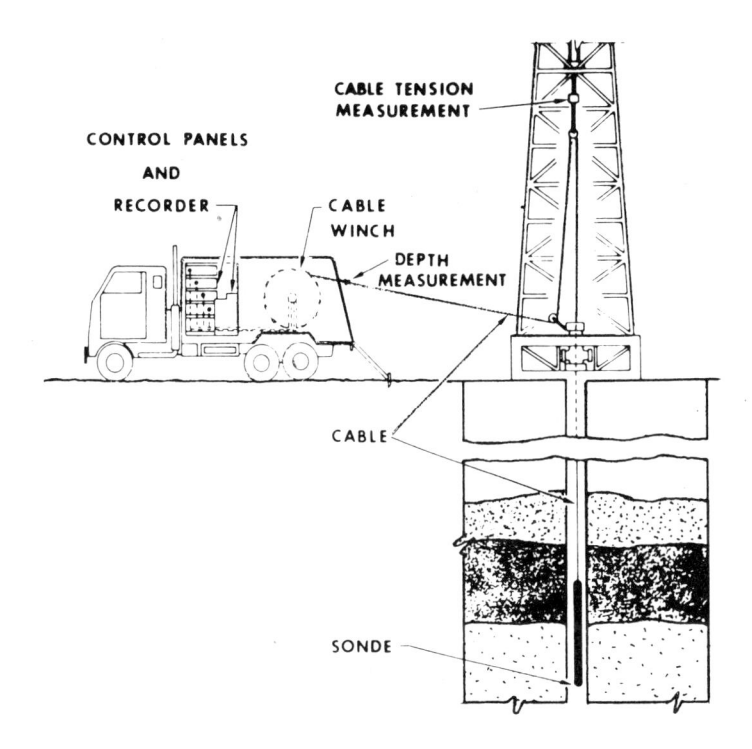

Arrangement of equipment for electric logging.

For example, five logs are placed side by side on a table and the engineer watches a formation at 5,800 feet. In four of the wells, the top of the formation runs from 5,792 feet to 5,808 feet. But in the fifth well log the top of the zone dropped to 5,940 feet. The engineer then checks the ground elevations on all five well logs to see that these elevations are all close. If they are, a fault has been cut — the result of the top of the formation having dropped around 140 feet.

Some faults trap oil; others are too low, and therefore in water, with no oil or gas. All oil production zones drift in and out of an area and then disappear, because of the way the earth is structured. Some production zones are sandy, some clay, some chalk, some limestone, and so on. Along the Texas Gulf Coast, some production zones bring sand with the oil and gas. This creates problems. The Austin Chalk formation seems to swell and seal off when a foreign

LOOK BEFORE YOU LEAP

water hits it. Some clays have a tendency to gum together if the wrong chemicals are left on the formation too long. Some of the formations are so hard and tight a frac job must be put on the well in order to get any production in an economical manner.

One of the best mineral leases to obtain is one that is a direct offset of good producing wells where it has been proven possible to drill through several known production zones.

The reason we explain a little of the geology is because an investor needs to have some understanding of what the promoter is talking about. In fact, you may have to know more than he does.

Remember the 300-acre lease that didn't look promising? The promoter and the engineer persuaded people to invest in the well because they said they knew the area — a good selling point. This can also be used as an example to help an investor. Ask the promoter if the geologist has experience in that area. And how much experience? Do the engineers have experience drilling and completing wells in the area? How many known production zones will be drilled through? What is the average well success in the area? Is this certain area badly cut up by faults? If the well has to be fraced, what type of frac* job is put on the well? Do these production zones normally carry a lot of water? How far away is the nearest documented production? The key word in this last question is *documented*.

Investing Time Before Investing Money

The time has come to invest in an oil and gas venture. The promoter arrives to show the investor a drilling prospectus. He will tell the investor where this lease is located and present some pertinent information. If you are the investor, you will be supplied with a small surface map marking the mineral lease, the surrounding producing well, faults, and the calculated depth of the proposed production zones to be encountered.

The Mineral Lease

The promoter will tell the investor, "This well we are offsetting is making a tremendous amount of oil and gas, just as this adjoining well over here is doing." And before he is finished, he will have covered the status of the surrounding wells. As a potential investor, you must inquire where he obtained his figures. The investor needs documented figures. (In Texas, the actual production figures from each producing well are available through the Railroad Commission in Austin. These records are public records.)

At one time, there was a particular oil company selling working interest in a well in Gonzales County, Texas. The lease was a direct offset to a producing oil and gas well. On the map with the drilling prospectus, the geologist showed the producing well to be making 1,500 barrels of oil a day. I had my doubts, because I had not been supplied with any production figures whatsoever. I visited the well many times while out in the field. I inquired from the geologist where he had obtained his figures, and he told me from a production engineer. I told him the figures were inaccurate. The geologist in turn asked me if I knew how much oil and gas the well was making a day. I said that I would guess it to be around 600 barrels, due to their production facility hookup. The company had an oil and water separator that would handle about 800 barrels a day and three 400 barrel storage tanks. The three tanks could hold only 1,200 barrels of oil before running over, and the separator had to be large enough to let the liquid have enough residence time to separate itself. In this manner, the oil will not go with the water, and the water will not pass into the oil tank. Another thing to consider in oil storage is that sufficient storage should be available for four days production, or possibly six days. The main reason for this is the problems of oil hauling. If the well had been producing 1,500 barrels of oil a day, the company would need fifteen 400 barrel tanks, or ten 600 barrel tanks, plus a larger separator.

I waited a few days. Then, I contacted the Railroad Commission and requested the correct figures. The well was actually producing 560 barrels of oil a day. The geologist

LOOK BEFORE YOU LEAP

did not want to know that information. He did get a well drilled, though. Guess what? Not a drop of oil. It was all gas production, but there was no gas pipeline nearby. Two miles of pipeline had to be laid for a sales line.

A word to the wise: Research all the facts and be sure they are documented. Ask the promoter to leave his material with you. Then begin your own investigation.

Chapter 2

SMALL COMPANIES AND PRIVATE CORPORATIONS

The potential investor in oil has a variety of kinds of companies to choose from — everything from the independent wildcat promoter to the big, major oil companies on the stock exchange. Good investments exist at all levels of the oil and gas business, but risks exist as well. We will not concern ourselves here with the major companies, whose exploits are closely monitored by the public press as well as by stock brokerage firms and stockholders.

It is with the small companies and private corporations (many of them reputable and good potential investments) that I wish to offer advice upon, for it is with them that the wise investor will need to utilize good judgement.

Behind the Public Relations

Unless an investor sees a test result on an oil or gas well conducted by a well tester or a government agency, I would not put much faith in a summary written by another company employee. Some companies hire people for publicity and promotion, and the fact that oil and gas comes from the ground may be the total extent of their knowledge.

I recall one particular well that had been drilled and completed on land owned by an attorney. I sat with him in his office one morning. He proudly announced that the oil

LOOK BEFORE YOU LEAP

well on his place looked good. It was producing 640 barrels of oil a day, and gas was tapped along with the oil. I assured him it wasn't producing that much oil and asked where he got his information. The attorney said he read about his well in the *Wall Street Journal*.

I said the report was inaccurate since we were still in the process of cleaning the well up and had yet to run any tests. The well was making about 200 barrels of oil a day, and the remaining fluid was water. At the time, we did not know how much gas the well was producing because we were not yet metering the gas.

Of course, the attorney was upset about the false report. When I checked into the situation, I discovered the report was taken from a daily write-up by the company reporter. It stated: *cleaning up well — 640 barrels of liquid recovered: 204 barrels of pipeline oil; 436 barrels of frac water; flowing pressure, 960 pounds and good gas flow with the liquids*.

The promotion writer took the total of 640 barrels of *liquid* and substituted 640 barrels of oil. Definitely a mistake! On purpose? Maybe not. The article was cleverly written, and it accomplished its purpose — it made the company look good.

Small Oil Companies on the Stock Exchange

To outsiders, the operation of the smaller oil companies that are listed on the Stock Exchange will seem odd. It *is* strange, but it *is* interesting, and in many ways it *is* funny. I know of one Houston based oil company, which had a good deal of money in the bank, but started raising prices. People within the company made side deals difficult to detect. Others set up their own companies, giving out large amounts of stock on credit like dispensing Dixie cups of water on humid days. They issued pumped up oil well-production reports. Of further interest is the fact that one company on the Stock Exchange formed three small companies to soak away from it substantial profits.

Stockbrokers have no way of knowing all happenings

Small Companies and Private Corporations

that take place within a listed Stock Exchange oil company. Brokers may believe they know and tell you they know, but they operate in ignorance of the company's day-to-day operations. I can report that one company in Houston, Texas, had seventy-five percent of the work force in the dark, ignorant of activities taking place right under their noses. Of course, some employees would have little reason to know the varied aspects of money flow and accounting practices of the company, being unconcerned with invoices, cash controls, land, or company stock issues. It is reasonable to assume that if three-quarters of the employees have no knowledge of internal happenings within an oil company, then stockbrokers in offices far removed from the company have little knowledge or experience with that company on which to base a sound financial opinion. The broker could guess or operate on judgments made on hearsay and the published report of the company, or statements of local writers or journalists. The writer or journalist, a hired, salaried person doing the daily job of reporting facts, could be controlled by the president or the chairman of the board and could create favorable texts of the company's operations, expenses, and financial returns.

All small oil companies operate in a manner that I am describing here. What can happen in the future has happened in the past. It would take months or years to investigate and study the different happenings that take place daily within a small oil company that covets being listed on the Stock Exchanges.

Employee Stock Options

Let us examine the matter of companies that sell company stock to its employees. Most stockholding companies have a stock purchase plan available to a select group of officers and top level management. A few companies have a stock purchase plan available to all

LOOK BEFORE YOU LEAP

employees. Some of these stock purchase plans are excellent and some are just "plans." Some purchase plans have little worth.

When I was with Florida Gas Transmission Company I participated in a very good stock purchase plan. In reality, the plan was a savings plan for the employees. The way the plan was established was a decided help to the company in a small way, and it greatly aided the employees in good practices of thrift. An employee was encouraged to sign to invest five percent of his salary each month in company stock. The company handled all the paperwork, and it set aside certain shares of stock to give the employee when the employee had participated in the plan for several years. In general this type of stock plan is a good savings plan, offering rewards for the years of service an employee gives a company. If an employee stays with the company for seven, eight, or ten years, he gains a firm financial interest in the company.

However, one oil company I worked with had a stock option program that I did not like, and I chose not to participate in it. This particular stock option plan had employees within a high level management group of the company buy stock at a fifteen percent discount. This plan was available to this limited group and only on particular offerings. Employees who purchased the discounted stock had to hold the stock for at least one year. With the fifteen percent discount, a stock selling for $20 a share on the exchanges could be purchased by this small group of employees for $17.

Once, within a six-week period, I was offered $45,000 worth of stock on one offering, $22,000 worth of stock on another offering, and $21,600 worth of stock on a third offering. Under this plan the higher level of management you had attained within the company the larger the amount of stock that was made available for purchase at the fifteen percent discount.

The stock could be purchased by cash payment or, having made a minimal payment, by monthly deductions from

Small Companies and Private Corporations

salary. I do not know of anyone choosing the purchase of this discounted stock on a cash basis. If an officer or executive thought the purchase price of the stock could hold for a year, or perhaps gain in price on the stock exchange, purchase on credit was the preference. If the stock exchange price held steady, the purchaser would make at least fifteen percent interest on the money invested in the stock. It is important that the investor gain insight to the magnitude of the 15 percent discount stock-purchase option.

I recall there were 22 employees within the company who could participate in this program; a number of these people were of higher rank than I was and they were offered a larger number of shares. However, I will use the nearly $100,000 worth of shares offered to me to explain that, with 22 employees participating during the six-week period in the offering, the total stock purchase was valued at $2,200,000. The 15 percent discount on this amount is $330,000, and the stock purchase on the totals was $1,870,000 — on credit. Let us assume 10 percent as a downpayment on the purchase price of the stock — a total of $187,000 cash was received from the 22 employees. This means the $1,683,000 is to be paid by the employees for the full purchase price of the stock. At the time of these offerings ($2,200,000) the oil company's stock was selling on the stock exchanges for about $20 per share, but selling to the employees for $17 a share — on credit.

Let us assume all 22 employees participated in the program. Each paid 10 percent down and agreed to have $1,000 deducted from their paychecks each month, until the balance owed was cleared. Therefore, a total of $1,683,000 would have to be paid on the balance to obtain title to the 110,000 shares of stock. If all 22 persons had $1,000 deducted from salary as payment toward the indebtedness, a monthly total of $22,000 was paid to the company for the balance owed. This means it would take the employees 6.375 years (76.5 months) to pay for the 110,000 shares of stock discounted at 15 percent. A nice help in the cash flow!

What would happen at the end of a year if the stock price

LOOK BEFORE YOU LEAP

dropped from $20 to $1 per share. Where do the employees stand? Who needs the money: the company or the employees?

There are two ways the small oil company will use this to its advantage. First, it may go to a lending company or a large corporation interested in investing in the oil business. To the lending company, the chairman of the board of the small oil company can suggest: "We have $1,683,000 to be paid in on stock, and we need to borrow $1,000,000. We will apply the payments directly to you, the lending company, until we are debt free."

The second and more advantageous procedure is for the board of directors to tell a big money investor: "We have just sold 110,000 shares of stock." A disclosure statement indicates 22 persons purchased the 110,000 shares.

The fact that the stocks were purchased on credit is not mentioned. To a company that may be planning to invest five or ten million dollars, this would, on the surface, look impressive. A great sales point, that is, if *all* the facts are not brought to the surface.

Abuses of Stockholders Money

Too, there is the aspect of escalated prices, and a stockholding company that purchases items from certain companies even if the purchase prices are somewhat higher. This does happen, and there is a reason.

For instance, consider the small company listed on the Stock Exchange that hasn't been in the oil and gas business for very long. One member of the board of directors of this small company owns a large interest in another company that sells wellheads (christmas trees*). And there is another member of the board who owns a hefty interest in a company that sells pipe. Another board member owns an interest in a small mud company. Now you can well imagine where these items will be purchased, even if they are higher in price than other suppliers.

Small Companies and Private Corporations

No cause for alarm, right? These three board members will be quite satisfied with the purchases going through a company that each owns in whole or in part. Every so often, though, a wellhead or the pipe or the drilling mud may accidentally be purchased from another company not associated with board-member owned firms. I've been reprimanded more than once for buying a product or service at a cheaper price, but from the wrong company. Are the stockholders' interests being taken care of this way? Definitely not!

Money saved, money earned! Maybe yes, maybe no!

Situations such as people making their own deals inside a stockholding company, or employees starting up their own company with the aid of the stockholding company's employees, could lead to a misuse of the stockholders' money.

Take into consideration another case. I am acquainted with the vice president of exploration of a stockholding company whose brother is in the land business. The brother's land business owned a company that put together drilling leases. In turn, he brought the leases to the land department of the stockholding company, which then accepted the mineral leases. You can guess who received an overriding royalty interest.

Another situation of interest! There is a man with a stockholding company who set up his own resources firm. He researched land records, production records, and division order records of the company. He then picked out what he wished to purchase and contacted the right people. He used employees of the stockholding company's land department, the production department, and division order department to prepare all his paperwork for purchase of interests in different properties, having no advantage for the stockholding company. Thousands of dollars were spent on an outside activity.

Consider the stockholding company listed on the Exchange that has within its organization structure three smaller companies. This small oil company has an

LOOK BEFORE YOU LEAP

operating company, a drilling company, and a little company in another state that does dirt work, furnishes frac tanks, and closes mud pits.

The operating company will obtain the mineral lease, promote the drilling ventures, have the wells drilled and completed, and then sell the products. The drilling company has the drilling rigs and drills the wells. The company located in another state does location and road work, furnishes frac tanks for frac jobs, then cleans the location after the well is drilled and completed, or plugged and abandoned.

The operating company obtains a mineral lease, and the marketing people sell the venture. The stockholding company obtains a drilling rig from *its* drilling company. The rig to be used on this well should be charging $7,200 a day plus fuel, but they charge $8,200 a day plus fuel. Where did the extra thousand dollars a day go? To the drilling company.

At that rate, every 30 days, $30,000 excess charges will go to just one rig. Multiply that by ten rigs, and you have $300,000 overcharge in a month. So, the parent company spends excess money on another company.

Excess money is also paid to the dirt contractors owned by the parent company for the closing of mud pits and cleaning up locations. When I was employed by one company, there were several mud pits and locations to be cleaned. I sent out bid notices to three different dirt contracting firms, one of which was owned by the company I was associated with. And, as it happened, it was that particular firm that submitted the highest bid. Using the low-bidding company, I gave the approval for work to begin.

It wasn't long before I received word that the president of the stockholding company, my employer, wanted work stopped immediately. Hire the dirt contractor owned by our company! Our contractor's bid was too high, I said. I was ordered to inform our contractor of the low bid, then to request submission of new bids!

Naturally, our company-owned dirt contractor got the

Small Companies and Private Corporations

work. (I might add, however, that I never saw the invoices for the completed work. I assume they went directly to the president of the company.)

Once, while working for this company, we drilled and completed an oil well in the state of Arkansas. The engineers used frac tanks to hold the oil from the well while cleaning up, which were supplied by the dirt contractor. The contractor sent one of his men to the well to pick up one of the frac tanks. At the time it was full of oil, and the contractor's man opened a valve on the tank and dumped all the oil in the mud pit and departed with the tank.

Soon after, it began to rain heavily. In time, the mud pit ran over. The oil oozed out of the pits first because it floated to the top. Oil seeped down a small hill, through a pine forest, and into a small creek.

The parent stockholding company paid $32,000 in damages and the cost of clean-up. In the process, this wanton carelessness lost them 400 barrels of oil. State officials were ready to kick us all out, and with good cause. The dirt contractor owned by the stockholding company caused the problem but never paid one dime. I might add, the landowners were angry at the company, although it had accepted the responsibility for the accident. It looked as if we had purposely dumped the oil and caused the environmental loss and damages.

How Small Successful Companies Operate

To be successful in any business, the person who starts the busness must know the field from the ground up. This is especially true in the oil and gas business, as each lease is different and each well is different. They may look alike, but they are not.

There is one small independent operator in Houston who has been very successful. Why? In talking to him, I found out that he set guidelines for himself — guidelines for everything related to his business — and he follows them strictly.

LOOK BEFORE YOU LEAP

1) He decided that he would not drill a well or take a property if he was short of money, even if the drilling venture looked really promising. By doing this, he stayed out of debt.

2) He would not take a drilling prospect unless he would be the operator. This decision insured that he would remain in charge of all operations, so that he could accept all responsibility. The load was on his shoulders, good or bad.

3) He classified his investors as big, medium, and small; and he made a point of contacting them before he took on a drilling venture. By classifying his investors according to the amount of money they would be investing, he kept a control on the costs and expenditures of each drilling venture. Big investors were used on deeper wells, and the small investors on shallow wells. By controlling drilling costs in this manner, his smaller investors could have the same percentage interest in their shallow wells as the big investors had in the deeper wells.

4) A very important decision involved his approach to drilling ventures that were brought to him. He would take an option to hold the venture for several days, during which time he would take the venture to experts for their opinions. He would listen carefully to any negative aspects the experts would bring up, keeping notes on everything. He did not trust entirely to his own judgement. (Taking the option was an expression of his own positive response, and he set out looking for anything bad that might override his impressions.)

5) Once he has determined that a drilling venture indeed looks good, he begins calling investors to show them the deal. What he gives them is a cost estimate with 10 to 15 percent added for error. (If he is unsure of all the existing problems, he will add as much as 20 percent for cost overrun.)

6) When he has all his investors signed up, he keeps them informed on operations on a daily basis, telling them everything happening, good or bad. He stays in close contact with them at all times, sending written reports and calling them.

Small Companies and Private Corporations

7) Each investor is given a statement of total expenses and total sales each month, or as soon as the information is collected.

8) Unless an investor specifically asks him to take the money in advance, he has a rule of not taking investors' money until it is needed.

At one time this oilman had a well in southwest Texas that came very near a blowout. With numerous problems, there was a cost overrun. However, he kept his investors because of the way he ran his business. His office is still open; he is still drilling wells; and he is keeping his investors happy.

Private Corporations

Some small companies and private corporations are set up with stock certificates that are not sold on the stock exchange. There are both good and bad corporations. There are pitfalls in almost any venture. I offer this as an example of just one bad thing that can happen. There are other ways investors have lost money in private corporations.

One private corporation I know of was founded in the late 1940s. Its offices were located in a prestigious bank building, and it began operations like any other small oil firm.

This company started drilling and completing several producing gas wells. Shortly thereafter, the company started going astray. Rumors had it that they were pulling off some quick scams. When I heard the stories, I talked to an attorney who had worked for the company. Although he did not tell me all of the trouble they had gotten into, I obtained enough information to pass on here.

This private company was receiving income from completed wells when someone from within figured out a fast money-making idea. When the trouble started, the company had more than 15 good, producing gas wells, with money coming in on their original investment.

The corporation had connections in New York — people

LOOK BEFORE YOU LEAP

with lots of money. Participating executives voted to print some stock and sell it in New York through these connections. Although they were not listed to sell on the stock exchanges, it was still proving to be a successful scheme. Their luck was due mainly to the fact that the company was sending a portfolio along with the stock, explaining how many wells they had producing and actually showing the original investment. Big money people began buying the stock. It was mostly through word of mouth that the stock kept in demand, and then more would have to be printed. Unfortunately, by talking enthusiastically, the first investors helped sell the stock for the corporation.

Before too long things started getting out of hand. Stock was selling like mad, but the investors received nothing in return. The company men printed more stock certificates, let them dry, and sold them. A small percentage of this new money was paid as dividends to some of the older New York stockholders. Soon it was time to print more stock and give the older stockholders 10 cents of each $1.00 brought in on these new stock sales. It was then that someone in the oil corporation got another idea.

The company picked up some cheap mineral leases. They cemented wellheads, or christmas trees,* on these cheap leases, but drilled no holes; they merely set the wellheads in cement on the ground. They snapped a few pictures and sent these along with new stock certificates being printed and sold. No gas wells, but definitely pretty pictures. Investors who had been buying the stock told all their friends about these new oil and gas wells this company had completed. No gas was being sold as yet because they were not tied into a sales line, they told one another. But, there were pictures of the wells!

The corporation's stock was selling so quickly that the man, a bagman representing the oil company in New York, would return to Texas with a suitcase filled with cash. A company man, meeting him at the airport in Texas, would attempt to guess the amount of money in the suitcase by the angle at which the New York man was leaning.

Small Companies and Private Corporations

Now the New York man who was selling the certificates directly to investors got a brainstorm. He planned one of his trips to Texas to include a few investors and prospective investors. The man had a gauge installed on each wellhead (where, in fact, there was no well at all). In the installation process, he set the gauge hand with a screwdriver so as to indicate pressure on the wellhead. Some of the gauges were made to read 2,000 pounds, some read 2,400 pounds and some read 1,900 pounds. If even I were to walk up to the wellhead and see this on top of the structure, I would believe the pressure indication reading on the gauge. The people coming from New York were supposedly to have a very successful tour; the plan was the best.

The arranged tour collected a planeload of people to look over this project. On arrival in Texas, the planeload of people were transferred to a bus for the grand tour of their "oilfields." The company planned a barbecue, Texas style, under trees near one of the fake wellheads. Fancy drinks and beer for all.

Now, we come to the fascinating segment of this scheme. At one of the wellheads, cemented in the ground where a well had never been drilled, the company installed a 30-foot piece of eight-inch pipe. This 30 foot pipe was hooked to the wellhead with half-inch pipe and with valves, and filled with crude oil.

The bus arrived on the scene so the stock investors and guests could view this one wellhead. Until now, no other wellhead valve was opened because, of course, there was nothing there. The investors and guests were informed they would receive a small *something*, a memento to take back with them to New York.

Shortly after the bus arrived at this wellhead, a pickup truck drove up. One of the corporation executives asked the driver (another employee involved) if he had brought the sample bottles. Then the executive proceeded to fill these bottles with the crude oil out of the half-inch line from the wellhead. Each guest was given a small bottle of crude oil for his very own, and each bottle was labeled: "'Texas Crude Oil — Well No. 50.''

LOOK BEFORE YOU LEAP

I am inquisitive by nature, so I asked the attorney why, if the well was supposed to be making oil, wasn't it hooked up to the production facilities and tanks. He confided that the driver bringing the sample bottles, arrived at the well under the assumption that he was to talk about hooking the well up to the facilities. Of course, this scheme was thoroughly planned beforehand and this answer would, more than likely, satisfy any unknowledgeable investor.

Now then, lets see how even this most superb promotion scheme backfired! Of course, the highlight of the bus tour was the Texas barbecue with beer and drinks under the trees, near another wellhead, one merely cemented in the ground. The investors held their crude oil gift (the bait) and were now moving to another area — a grand finale under a hot Texas sun. A few drinks will help even the worst sales pitch, but it did wonders for the big event about to take place. It was suggested by someone that photos be taken for souvenirs.

The corporation had another wellhead rigged near the barbecue pits. A large propane tank tucked in an out-of-the-way area was hooked to this wellhead with a half-inch line. When the investors were ready to eat, one of the corporation people told the truck driver to turn the well on and light the flare line. He went to the wellhead, opened a valve connected to a line running to a pit, and lighted the propane coming from the line. A huge flame was burning, leading the investors to believe the well was flowing gas to the air. More pictures of the investors were taken by the well that supposedly burned natural gas. Still another memento to take home. The sales pitch over false schemes worked, and once again the corporation printed stock madly, selling it as soon as the ink dried. However, before they could sell all and run, trouble arrived.

A driver for a trucking firm attempting to deliver a piece of oilfield equipment got lost. He wandered onto one of this private corporation's leases, where a cemented wellhead stood atop the ground.

The truck driver, realizing he was lost, began to turn around his truck. In turning he became stuck in mud. The truck had a winch, as do all oilfield trucks. The only thing

—38—

Small Companies and Private Corporations

he could find to tie the winch line on, to pull the vehicle out of the mud, was one of the wellheads. He engaged the winch and started to pull the truck out. No such luck! The dummy wellhead was pulled up and turned over. The cat, so to speak, was out of the bag.

Take care in investing in private corporation stocks. Investigate the company and its principals and take your *own* personal tour (not a free food and beer tour). Examine so called moneymaking ideas. Spend a little money investigating. It may save thousands, even millions of dollars.

By June, 1983, I had read and heard the reports of the swindles in the state of Utah. These swindles involved members of a church affiliated group. There was over $100,000,000 (perhaps closer to $200,000,000) lost. People invested money with a promise of unheard of profits. It is my understanding that the church group asked the state of Utah to take action against the swindlers. No matter what state one lives in, it has no influence over personal investments. No one will look out for your investments unless paid to do so.

Let's say I know a stockbroker whom I call, as well, a friend. And I own a small moneymaking organization. Now the time has come for me to set my sights on better things, and I want to make big money and retire from the business. Keeping in mind that buying and selling stocks for other people is the broker's occupation, I tell him that I have 100,000 shares of stock that I want to sell for $20 per share. It probably won't make much profit, but I agree to give him two dollars a share to sell it.

The stockbroker will receive $200,000 from me, while the investors are going to give him $30,000 (which is 1½ percent sales commission). My sales payment to him is almost three times what he will receive as purchase commissions from investors. Meanwhile, I have sold the stock and made $1,800,000 and "retired."

This sort of thing happens all too frequently. Don't pass up an opportunity to make money in the oil business, but be wary of schemes such as this. Do your homework.

Part II

DRILLING PROSPECTS

Chapter 3
WORKING INTEREST IN A DRILLING VENTURE

Before you commit your money to a drilling venture, there are several questions you *must* have answered:

1) What percent of the revenue comes back to the working interest? If it is below 70 percent, investigate thoroughly. There are some highly productive drilling areas that may only have 65 percent to 68 percent left for revenue to the investors. This could be because the wells will produce a large daily volume, and that helps the investor receive better returns.

2) How much working interest will the oil company keep? As a useful guideline, if the oil company thinks the drilling prospect is good, the company will try to take as much working interest as they can possibly stand, although this sometimes can bankrupt even a good company. Once, I observed a small oil company drill and complete five good wells. However, they were drilled to a T.D. (Total Depth) of 13,000 feet. The company was forced to put all of its reserve capital into drilling these wells. It took a long time to hook up the wells to a sales line, and the cash flow return was slow. The small company had to sell good producing wells. I can add, these people knew the drilling business; they completed good producing wells, and the investors have been receiving returns now for six years.

3) Can you, as an investor, request from the promoter a list of other potential investors in this drilling venture? Yes! The investors can help each other. Meetings or

DRILLING PROSPECTS

conversations among people with mutual goals often produce excellent results. And this tends to get all wheels going in the same direction. Compounded energies and knowledge may also keep investors from making mistakes.

Giving Away More Than You've Got

I can describe one kind of mistake that has taken place from time to time, one I know from personal involvement. This happened many years ago when I had very little knowledge of how to find investors or how to put a drilling venture together. I personally obtained a 400 acre lease. It was a good-looking oil, gas, and mineral lease. This acreage had producing wells on two sides and it had the additional advantage of a gas sales line passing across it.

I acquired the lease using the drilling contract method. Also, I knew the person who owned the mineral rights on the property. I approached this owner, and I was forthright in admitting that I didn't have money, but I thought a good producing well could be made on this land. I asked for a six-month drilling contract; he allowed me a year free of charge. While attempting to set up the drilling operation, I was advised by someone to have my venture reviewed by a particular geologist.

The geologist enthusiastically examined the prospect and said he would have it drilled and give me one-eighth (1/8) *overriding royalty interest*, which was 12.5 percent. Now, the leaseholder already had one-eighth (1/8) interest. I explained to the geologist that I wanted to be informed when a drilling rig was moved onto the site, that I wanted to be on the well site when the open hole log was run. He agreed to the idea, although we put nothing in writing between us.

One day, I decided to go look at the leased property. It was a big deal for me at the time because it was my first well to really be wholly involved with. I arrived at the lease and saw the drilling rig was being moved out. I found the toolpusher and began to ask questions, requesting the open hole logs be made available to me. I was informed that the

—44—

Working Interest in a Drilling Venture

geologist had them. When I called him, asking for a copy of the logs, he told me they had been thrown away. *It was a very dry hole.*

At that time, there were just two well logging companies in the area. By luck, the first one I called had logged the well. I got out my copy of the mineral lease, proof of my interest in this well, and went to the logging company office. When studying their copy of the log, I found verifications that there were two good production zones. One of the zones we had looked for. However, the second zone was new. After numerous attempts to try to locate the geologist, I finally found him at home one evening.

I explained to him that I had verified a well could have been made. He stuttered for a few minutes and then admitted that he was in a mess. He confided that his secretary had made a mistake, and he — through her error — had sold thirty 1/16ths interest in the well. (Keep in mind, a secretary should do only what she is told to do.) The geologist only had 75 percent interest to sell in the first place. He had sold almost 200 percent or two wells in one hole. There was no way he could complete the well.

This experience was good for me and the leaseholder. That deal helped me to learn to cull honest people from the dishonest. I wrote a letter and took it along with the well log and mineral lease back to the leaseholder, a retired judge. After explaining the turn of events to him, he told me it would be pointless to sue the geologist.

Today there are two good wells on this lease, and 22 other producing wells surround the lease. An attorney friend tells me the 400-acre lease that I had was the best of all the surrounding ones. The leaseholder's grandchildren received one-sixth (1/6th) interest plus an overriding royalty interest and bonus. So there was a profit after all. But this doesn't always have a happy ending. So, as an investor, you must check all interest details. What percent working interest is for sale, and who else may be investing?

That geologist is not in the oil and gas business today; he's teaching school. Six months after this event I discovered that a drilling mud company had furnished the mud for this well, for a paid-up interest in the well.

DRILLING PROSPECTS

Improving Someone Else's Property

Another situation I might caution against is drilling a well that improves someone else's property or mineral lease, without giving a benefit to the investor.

As an investor, assume you took an interest in an original lease of 500 acres for one well. The promoter sells you interest in a proposed well that is placed in the corner of a piece of land farthest from producing wells that are situated on three sides of this mineral lease. These producing wells are drilled to a total depth of 7,600 feet. On each of these three sides are different companies holding mineral leases with good, productive wells, and on the fourth side of the lease, a major oil company holds the rights to 1,000 undrilled acres. The promoter attempts to sell you an interest in a 10,000-foot well and tells you about a good production zone at about 7,520 feet; and still another zone at 9,860 feet. In a roundabout way this makes sense. If there exists the possibility of making a well at 7,520 feet and warranted proof is provided, go deeper and see what will develop.

Now we come to the catch! The oil company (or the promoter) has already talked to all of the four companies involved in the surrounding area. Three of the companies have already drilled down to 7,600 feet and have told the promoter that they each will give him $10,000 for all the information he can get on the well drilled to 10,000 feet. Next, the fourth company offers the promoter $50,000 for all the information obtained while drilling the well. Now $80,000 has been offered to the promoter for information on the well you are about to invest in. You may not know it, but you've been taken!

How do other companies improve their leases, and why is it cheap for the money paid out? The oil company is on the fourth side of the lease and it is paying $50,000 for information that will either prove its lease productive, or not. (Production zones aren't always good, and just as easily new production zones might show up.) This money paid out will give the oil company a look into the earth down

Working Interest in a Drilling Venture

to a total depth of 10,000 feet plus other pertinent data without drilling for themselves what might prove to be a dry hole down to 10,000 feet. The other three companies on the surrounding sides may obtain information for the 2,400 feet, without having to drill their well to the T.D. of 10,000 feet.

The 500-acre lease will carry three wells with a spacing of 160 acres. The promoter is trying to sell you the one well that is farthest from the offsetting productive wells. Assuming he gets his well drilled and it proves profitable, you have just paid money to prove two more producing wells may be drilled and completed between your well and the already producing one, while providing him with knowledge of a new production zone below 7,600 feet. You will probably never get an opportunity to invest in the other two wells unless you tell the promoter that you want an option to buy interst in any other wells that may be drilled on his 500-acre lease. That is, if you do buy an interest in the first well.

If the promoter completes a well in the first hole drilled, he has zeroed in on two definite and excellent locations to drill. He will try to hold these other two well sites back and invest as much of his company's own money as possible.

Why did the three companies with producing wells pay money for information in the well from 7,600 feet down to 10,000 feet? In this case, they found out there was a possibility of finding another production zone below 7,680 feet. When their wells deplete or become uneconomical to produce, they may drill down to 10,000 feet before plugging and leaving. They would only have to drill out 2,400 feet and, if production is found, set a 2,460 foot liner and have a producing well again.

As an investor, you should ask the promoter if any information is being sold. Why? Because it is information that you have paid for. Also, the money taken in from this information should go on the A.F.E.* to help cut the cost of the well to the investors.

Be positive. Gather all the facts about the *working interest* before you invest.

DRILLING PROSPECTS

Self-propelled well service unit. This unit is being employed to remove the sucker rods and tubing from a pumping oil well. A: Sucker rods hanging from a rock. B: Tubing standing in rock.

Reserves

Reserves are the amounts of oil and gas — projected or real — that may be in a specific production zone. Before drilling, calculations are done to attempt to establish the reserves in a production zone area. Calculated results may also establish the millions of cubic feet of gas that may be drained from a specified area. However, there is no economical way to recover *all* the oil and gas from a production zone.

Let's assume we have drilled a well and found an oil

Working Interest in a Drilling Venture

bearing zone, and all indications point to the possibility that it will produce profitably. We set casing, perforate and run production tubing in the well — a shallow well approximately 3,000 feet deep. The well begins to flow oil into the production facilities with an initial flowing pressure of 60 pounds and producing 40 barrels of oil per day, with very little water.

Forty barrels of oil at $30 a barrel is $1,200 a day. The *working interest* on this well is 74 percent. Those persons who invested in the working interest would receive $888 a day (minus taxes and operating expenses). If this well flows for one year, however, the flowing pressure will be down to zero, thereby producing five barrels of oil a day, plus eight barrels of water. Five barrels of oil a day at $30 a barrel will be $150 a day. The 74 percent working interest now receives $111 a day (minus taxes and operating expenses). The eight barrels of water a day that is being produced adds to the operating expense because the water must be disposed of. In a 30-day period, investors would be receiving $3,300, but suffer higher operating expenses. When the well stops flowing, a workover* rig is obtained and a down-hole pump is installed along with sucker rods* and a pumping unit. This creates more expenses to the costs of the original well.

The oil well now pumps fluid into the tank at a rate of 12 barrels of oil a day, plus 15 barrels of water. There is more water content now, and its disposal causes expenses to further escalate. Later, the production drops to 1.5 barrels of oil and 24 barrels of water a day. The operating expense has become greater than the revenue. Money is being wasted, so production must be terminated. The oil that is uneconomical to recover is left in the ground.

Companies that have several oil wells in one area will try water flooding. In water flood projects, water is pumped back into the production zone in one well with the hopes of forcing the oil into other well production zones. There have been as many successful water flood projects as there were unsuccessful ones.

The problems of draining the gas reserves will be the same as the ones draining oil reserves. Only the equipment for

DRILLING PROSPECTS

A medium-depth drilling rig.

the draining procedures will be different. First, all gas that is produced from a well will move into a sales line against pressure. All pipelines must hold a certain amount of pressure in order to move a specified volume of gas to the purchaser. That is, assuming the gas well has enough flowing pressure to produce sales gas to flow into the pipeline in an economical manner. Finally, the gas well will no longer flow into the sales line at an economical rate, and a gas compressor must be installed. This compressor may be purchased or rented, but it is an added expense. The function of the gas compressor is to suck the gas in at a low pressure and discharge it into the pipeline with enough pressure to increase the sales volume. We have increased the sales volume to an economical rate. However, salt water is flowing with the gas, and there is a compression expense

Working Interest in a Drilling Venture

and water disposal cost. Later, with gas pressure lowered on the suction side of the compressor, a larger compressor is installed, and more water is mixed with the gas. The operating expenses have become too great, and money is lost. The well must be plugged and the remaining gas left in the ground.

Depletion

The promoter should furnish investors with the potential oil or gas reserves and the actual amount of oil or gas that may be recovered. In the recovery or sale of the reserves, the investor is supplied with what we call a data sheet, illustrating the initial rate of production and then the declining scale. Ninety-nine percent of all wells fall off in production, rather than increasing in production, due to the fact that the reserves are consistently being depleted. In some cases, at the beginning, production may increase for a brief period of time, but then it will fall off again. Possibly, this is due to a water drive, another little pocket or new area opening up into the well zone as initial pressure drops. Investors need to receive the information as to the amount of reserves, and the amount expected to be recovered during the first year or two, and expected recoveries in the years thereafter. This information aids in calculating the cash flow from the proposed production. Remember, a production zone cannot be 100 percent pure oil or gas, and one cannot economically recover all reserves.

Also, just because oil is in the ground does not mean it can always be extracted. There exists one oil bearing zone in Central Texas that is carrying 60 to 80 percent oil. At this time, no one has found a way to get the oil out of the zone, as it is made up of shale with very little sand. The zone is 80 to 120 feet thick, and covers a fairly large area.

DRILLING PROSPECTS

A modern well logging operation.

Chapter 4
RE-ENTERING OLD HOLES AND WELLS

There are various new methods and procedures that have been brought into the oil and gas business that, on first view, seem difficult for a person outside the business to understand. The research and development in all phases of oil and gas production have cost literally billions of dollars. Let's not forget that mistakes as well as inexperience have been contributing factors to higher costs. People tend to learn from mistakes and varied experiences, not only from their own but from others' as well.

Supply and Demand

In the earlier years of oil and gas development, supply and demand were important considerations for determining if a well should be completed or if the hole should be plugged. Now, let's stop and think: supply and demand tend to control the price of what something should cost, correct? However, that is not the *supply and demand* that I refer to.

You might have heard someone say regarding a well plugged years ago something like the following: "They found oil or gas but they don't want anyone to know about it." Now, if an oil company found oil or gas that would produce in an economical manner, why would they plug it up and leave it?

Not too many years ago, natural gas was used as a fuel

DRILLING PROSPECTS

only in a small area situated near the producing well. Why? There were no major pipelines to transport gas to the consumers in distant cities, and there was insufficient demand to overcome transportation problems.

Years ago, a South Texas producer had to have millions of cubic feet of gas producing a day to get a gas contract, and he would nearly always have to lay his own sales pipeline and guarantee service 24 hours a day, 365 days a year. The price was extremely low, but if a gas producer could sell the gas, he was in business. So, if you were drilling for oil and found gas only, the well might, for economic reasons, be plugged.

Many of us are probably familiar with the statement: "I know where there are gas wells shut in, and the oil company will not produce them. They want more money for the gas." This is true in one respect. Major pipelines do not just say, "We've got the money, start working."

A study is made on all aspects of gas reserves, and then action is taken. First a pipeline company will try to lay a pipeline as straight as possible, taking into consideration the terrain, the gas supply, and the sales points.

Let's say I know where there is a large supply of gas in the south, and there is an unsupplied market in the north. I have a potential huge supply as well as a potential huge market, and all I need is millions of dollars. I organize this as I believe Tennessee Gas Transmission would do it.

Tennessee Gas Transmission is a reputable, well managed organization, one that didn't "just happen." It was a large scale, carefully planned project for the oil and gas business. Gas purchasers were put to work to locate gas fields, determine the amounts of the gas supply, and determine its price. Gas sales people were also put to work to canvas for consumers.

Of course, there is the D.O.T. (Department of Transportation) to consider. As interstate commerce is involved, this government department must be able to get its two cents in (two cents that will cost millions of dollars). Now, assuming all requirements are satisfied, a pipeline is laid. This is an original pipeline that is in business to transport natural gas to a market.

Re-entering Old Holes and Wells

Oil company A is drilling for oil and finds gas. However, they are located ten miles from a main gas pipeline. The pipeline people and the producer confer. There seems no way either of them can lay a pipeline for one gas well. The frustration is, the well looks promising. The alternative would be to leave the well shut in until the price goes up, or until more gas is discovered within this area, which is situated ten miles from the pipeline. (This is one reason why some wells are shut in; it's just considered good business.)

Let's look back at the oil well business several decades ago. There was no real market for natural gas before the 1950's. Wildcatters were (and are) people drilling for oil production where no possible production has been proven. Assume a wildcatter drills a well in a remote area. Years ago the drilling people were able to drill only shallow wells compared to the depths they drill today. A particular wildcatter drilled a shallow well and found two zones carrying oil. He tried the first zone, and it would flow only about 5 barrels of oil a day. That's not enough. He tried the second oil carrying zone, and this zone flowed 4 barrels a day; obviously not enough. The well was plugged, and the oil company left the area.

The reason for plugging that well was just plain everyday economics. The wildcatter wasn't hiding anything. Oil must be refined, and no refinery in the world has all of the oil wells from which they obtain crude oil standing in the refinery yard. The crude oil must be transported from well to refinery by pipeline, or by boat or truck. Although the wildcatter discovered the oil, the economics of producing the oil well was a problem. The nearby highways weren't developed to transport the oil, and trucks were few; in that area, the crude oil pipelines were rare. If the wildcatter was to produce at 5 barrels a day and could receive 50 cents a barrel, it would mean income of $2.50. The leaseholder receives 1/8 of $2.50, which is approximately 31 cents a day. This leaves the wildcatter $2.19 a day, providing there is no overriding royalty. In 30 days, the wildcatter had $65.70 worth of crude oil, minus expenses. The only solution: plug the hole and leave it.

I've not ventured onto a wrong track in alerting investors

DRILLING PROSPECTS

to examine drilling prospects and the entering of old holes or wells. A recent avenue has opened in the exploration of oil and gas that is highly profitable at this time. Many areas for finding commercial oil or gas have been overlooked by a large portion of small, independent oil companies and promoters. But not by some of the major oil companies.

I pointed out that Tennessee Gas Transmission set up a pipeline system to deliver gas to the northern United States from the fields in southern states. There are a number of companies in the pipeline business that follow this same south to north delivery system. For instance, in the late 1950's a pipeline was laid from the west to east, along the Gulf of Mexico's inner shoreline.

The state of Florida used imported butane and propane (in cannisters) until Florida Gas Transmission Company moved into the pipeline business and transported gas to the state. When this good company did that, it opened up a new area of exploration for the drilling people. A sales line was made available in a new territory for the drilling companies to wildcat* in the states of Alabama and Florida.

Old Well Logs

We've examined why many wells are left uncompleted, the hole plugged and abandoned, and gas wells left shut in. Therefore, the different approach to locating a good mineral lease is to search old well logs that had a good oil or gas show. We can even consider locating old well-drilling and testing reports. I know at least two major oil companies that have people sitting in log libraries consistently researching old well logs. (A Log Library is a company established to house any well log of a specific area that they may be able to obtain.) Persons looking at these old well logs are experts in their field and presumably know what they are looking for. Some people will start examining old logs of certain areas of past exploration because a story is circulated about a well test that looked good, or a driller on a

Re-entering Old Holes and Wells

rig has passed on some drilling stories.

At any rate, this is a good approach. Take, for instance, the wildcatter who drilled the shallow hole and tested four barrels of oil a day that once would only have created 50 cents per barrel. Today's prices, $30 a barrel for crude oil, makes a great deal of difference in the economic picture. Not only the difference in prices, but today's oil recoveries are so much more advanced. The well could possibly produce two, three, or even a hundred times more than it could have been made to produce some years ago, through new stimulation programs and well completion procedures.

To illustrate the soundness of this different approach and the money that can be realized, I can cite an instance of a particular oilfield of which I am familiar.

In 1944 and 1945, in a drilling operation, two major oil companies found 20 feet of oil-saturated pay zone at around 9,600 feet depth. At that time, they were using nitro to shoot the wells in an effort to obtain production. I read two well reports and was duly impressed. It was a lesson in the use of nitro. The reports read something like this:

> Shot well today with 1 quart of nitro. Well did not blow back. Started swabbing. Recovered 2 barrels good crude oil plus 5 gallons of water. Will shoot well with 1 3/4 quarts of nitro tomorrow. Next day. Shot well with 1 3/4 quarts of nitro. Well did not blow back. Started swabbing. Recovered 2 barrels good crude oil plus some water. Will shoot well with 4 quarts of nitro tomorrow. 3rd day. Shot well with 4 quarts nitro. Recovered 2 1/4 barrels of oil plus 1/4 barrel of water. Suggest plug this zone and come up hole to test other zones. If 6 to 7 quarts nitro does not bring enough oil, it's not any good.

That is what both companies did. They had almost identical results. Together, they drilled a total of 5 wells trying to get the oil out of the ground. The companies did what they could with the tools and the knowledge they had at that time. The production zones were just too tight to let the oil out.

DRILLING PROSPECTS

Two years ago, an engineer heard about the results the two major companies received from their 1944-45 drilling ventures. He secured the information available on these old holes and put together a large mineral lease. There are six producing oil wells in the area today. The minimum oil production is 224 barrels of oil a day, and the best well is producing 361 barrels a day. If we take the lowest producing well (224 barrels) at $30 per barrel, it pays a total of $6,720 a day. Assume 65 percent of this oil goes to the investors, i.e., $4,368 per day; at this rate, a 100-day total paid to the investors is $436,800. A well of this caliber costs around $860,000 to drill and complete. Without a doubt, this would be considered a profitable investment.

Not only are the crude oil price differences between the 1940s and the 1980s a great incentive for the study of old wells, but also new well-completion procedures make a difference.

New Technology

Getting oil out of a tight production zone is a procedure that was not in the oil industry's book of knowledge in 1944. Today it's a fact. We might add that currently more refined drilling technology is available to drill deeper into the ground than before.

If someone approaches you to invest in oil ventures and is using old well logs as an incentive to drill, examine the past findings carefully because it could be an excellent potential drilling venture. However, unless you make it your business to become an expert in deciphering well logs, I suggest you ask for the logs and take them to an expert and request assistance. For a sound venture of this type to be effective, knowledge must be invested with a great deal of time and study. Some people do not keep up with progress within their own trade. Some unknowledgeable people come into the oil and gas business from other professions. And those persons who have been in the oil and gas professions for any length of time have to look over new procedures. They may approve, but they will speculate and advise. They will

Re-entering Old Holes and Wells

seek advantages and disadvantages and while looking something over, they are learning.

There is a procedure that is referred to in the oilfield as *re-entry*. Personally, I have little faith in this procedure, but my opinion has no bearing on the economics of re-entering an old hole or old well.

Some persons might take an old log, find something they think looks good, and try to go back into the original hole, long since drilled and now plugged. One may buy this type of lease with the idea of going back into the old producing wells, instead of drilling new wells. If everything works properly, it is cheaper than drilling a new well. This is what the term *re-entry* means in relation to old holes and old wells.

Plugging Holes vs. Plugging Wells

However it is important to realize that plugging a hole that was drilled and *not completed* as a well is a different procedure from the plugging of a well that was *producing* or one for which there was an effort at producing. The drilling rig that drilled the hole will do the plugging on an uncompleted well, while a workover or completion rig* will usually plug an old well. There are companies that are set up just for plugging operations. Also, cement plugs are set to protect the fresh water sands and also retain any gas or water or oil more or less in its original position within the ground.

The plugging of a hole (an uncompleted well) is different. It is an open hole that is plugged; no production pipe or casing has been set within the hole. However, surface pipe would have been set as a protective measure. The drilling rig will set cement plugs in the open hole where specified and then move up the hole and set cement plugs in the surface pipe.

The plugging procedure involves pulling as much of the production casing out of the ground as possible. Retrieving part of the casing from the ground will save some of the

DRILLING PROSPECTS

money invested in the well. To retrieve a portion of the casing from the ground, a rig is set up over the well. The rig is then attached to the production casing by connection procedures; the pull on the production casing begins. Those persons performing this process must know the type of pipe in the ground. Knowing how much the pipe stretches per 100 feet and with what amount of force will enable the expert to successfully remove the pipe. This approved process determines where the pipe in the ground must be cut into in order to retrieve as much of it as possible. Sometimes this procedure creates a problem. A higher second cut may have to be made first, in order to retrieve a portion of the pipe. Then, most often, a cement plug is set inside the production casing that was left in the ground. Another plug is set in the open hole above the production casing, and then plugs are set in the surface pipe.

The Dangers of Re-Entry

Most of this explanation of plugging operations is of importance to the investor. No investor wants to lose his money. Someone may offer a cheap prospect on a re-entry program, and investors must be aware of the risk. Perhaps we should examine some of the dangers.

Let us say I find a good mineral lease, and in my study of it I find two old well logs with two potential oil production zones. The people who drilled the wells years ago had a good oil show; however, at that time, it was not profitable. I find some investors and propose to them that we re-enter an old hole because it would be cheaper than drilling a new well.

The first item on the agenda is to obtain a rig and begin an operation with a detailed outline. The cement plugs are drilled out of the surface pipe and then the first and second plugs in the open hole are drilled out. Trouble hits us on the second cement plug.

Since I made a study of this old hole, I know there is a 15-foot cement plug on the second open hole plug, and that's giving us problems. Exactly what is taking place?

Re-entering Old Holes and Wells

We drilled about a foot of cement on the second open hole plug when the drilling became easier. It did not feel as if we were drilling cement because we were moving too fast. (It's possible the drilling people neglected to set a full 15-foot cement plug, as they claimed they had done.) So, we drill a little more. After a short period of time, our drill bit begins to torque up; simply, the drill bit is more difficult to rotate in the hole. The drill pipe is pulled up the hole to circulate the drilling mud. Then, it's decided to ease back down and drill more. Our driller eases the drill pipe down the hole and begins operation again. But the torque on the drill pipe becomes an obstacle. The drillers decide to change bits, but the pipe will not move out of the hole.

We rent more equipment, and 5 days later the drill pipe and the bit are freed.

Then, after we get the lower section of the pipe with the drill bit attached, we discover we have slipped off the top of the second cement plug and are drilling to the side. This is called side-tracking.* It is impossible to drill this way because the hole will be crooked.

I have said previously that I have little faith in re-entering an old hole. I have worked on a rig and I know what takes place when one is being plugged.

1) "Men, we are going to move this rig as soon as we plug this hole," the driller orders. Rig hands can and will throw anything down the hole. It matters little to them; the hole is of no use any more. It's to be plugged and deserted and will not bother another living creature. Suppose a piece of iron is on top of the second cement plug and off to one side. Subsequently this would force a drill bit, going into the hole at a later time, to drift away from the iron; in other words, drill off to the side.

2) There is no way anyone can know how cement will set up in constancy of hardness 7,000 feet in the ground. At extreme depths there may be hard as well as soft spots and highly irregular patterns in the cement. This is mainly due to temperature and water saturation from the formation. In addition, mixing and pumping the cement down the hole while spotting the cement plug with drilling mud may be a

DRILLING PROSPECTS

factor. Some drilling mud may even become mixed with the cement plug.

Let's assume we are going to set a plug in an 8 7/8 inch hole. On the top of this plug, one third of the cement sets extra firm and two thirds of the top of the plug is soft. This condition would have a great tendency to let the drill bit drift off to the soft side of the plug. It would result in sidetracking and drilling a crooked hole.

Now look at the case of re-entering an old well that has been plugged and a portion of the pipe retrieved from the ground. We've said the plugging procedure is different on an old producing well with respect to the open-hole plugging procedure.

Suppose I obtain a good mineral lease that contained some old producing wells that had been plugged and abandoned. I make my investigation into the wells, and I find a potential producing zone that had been overlooked. I examine old well records and discover the exact depths of the cement plugs and the exact depth the production casing was cut off in the ground. Now, I'm information wealthy. All I need is my investors, then I'm ready to proceed.

We move a rig onto the leased location, with all of the other necessary equipment, and begin work. The cement plugs are drilled out of the surface pipe, and then the open hole plug, where the production casing had been pulled, is drilled out. No problems.

We work to get our drill bit to go inside the producton casing that was left in the ground. Unsuccessfully! Finally, two days later we get the bit inside the old casing and begin drilling the cement plug. It is fairly easy drilling for about six feet, but then the drilling slows. No depth is gained. After two hours of hard drilling, it seems to go easier, and we are drilling once again.

One of the roughnecks on the rig catches a few samples of the cuttings from the drilling mud. Before he tells anyone what the cuttings look like, the drill pipe starts to torque up slightly. The roughneck tells the driller he found a lot of metal and earth cuttings but very little cement. Drilling is halted, and we pull up the hole and circulate our drilling

Re-entering Old Holes and Wells

mud. We had side-tracked off the cement, cut through the side of the old production casing and drilled a new, crooked hole. In vain we work two more days in an effort to correct our problem. We have to shut down and plug the well, giving up the operation. All of our efforts and money are a total loss. It had looked like a cheap procedure to obtain a good producing well.

Picture if you will a soda straw in a glass of water (no ice). The straw will not stay upright nor will it stay in the middle of the glass. It leans over, against the rim of the glass. In most instances, that is exactly what can happen to the production casing when it is cut off in the ground with no means of support. The pipe has no reason to stand erect and will lay against the side wall of the hole. Therefore, when trying to enter the pipe straight down the hole with the casing leaning, the procedure clearly will be complicated. Not only that, if the old casing does not hold up fairly straight, the drill bit will cut the side of the old casing and start a new, crooked hole.

Now we can understand why a drilling method that may appear to be inexpensive may actually be costly to the investor. Luck has to play a big role in the success of many re-entry procedures. When you drill a new hole in the ground, you are the boss. You have the control factors, and you aren't dealing with so many unknowns.

A re-entry well operation can be expensive. One major company spent $1,300,000 on a re-entry and had to abandon the project. Their highly experienced people should have known better, but it was a deep hole and it looked like a money saving deal.

Roughnecks* who clean up the rig-site floor can throw things down the hole believing no one will ever return. That was what happened in the case of the company that spent over a million dollars on a proposed re-entry. It drilled inside the old production casing. The top of the cement plug inside the casing was 600 feet down from the open end and had been cut off. When they got to the top of the plug, they were unable to drill. The roughnecks would tear up a bit and then come out of the hole. When they tried to re-enter the

DRILLING PROSPECTS

pipe in the ground, they were stuck continuously.

Finally a junk basket was sent down to catch the samples of debris surrounding the drill. When the basket was brought up, it contained part of a pipe wrench and parts of a chain and pipe slips. The bit was replaced with a different cutting head, and drilling resumed. Metal was being cut, but then they slipped outside the pipe and became stuck. After all efforts to drill further were foiled, this company left part of their drill pipe and the small drill collars in the ground (and, in a sense, $1,300,000 remained in the hole).

Promoters and companies will always look at old dry holes and old wells for the possibility of re-entry. Those who may be approached to invest should be cautious and obtain all of the facts of the re-entry venture. Although I do not care to condone the idea of re-entry, I indulge myself in drilling prospects around old dry holes whenever the opportunity is there. The financial possibilities are limitless if surveys show it to be a healthy risk, and all evidence is presented as such.

I know of one man in Oklahoma who has his own drilling rig, and has been very successful in re-entering old holes. He has been in the drilling business for years, and he knows what he is doing and what tools to work with. Working from two considerations, he will look at any old well log in his home territory. The first concerns whether or not he will make a successful completion, and the second is the cost factor of re-entering versus drilling a new hole.

Re-entering has to be cheap, or he will not touch the venture.

This man told me about one well he re-entered that was very interesting. To drill and complete a new well would have cost an estimated $182,000. He determined that re-entry would cost no more than $90,000. How he arrived at this figure, I don't know, but he presented this to his investors as a stopping point in the venture. Should they reach the $90,000 in expenditures without having a successful re-entry, all work would be stopped and no more money would be spent.

He completed this well, and it was profitable, even though

Re-entering Old Holes and Wells

there were problems in re-entry. (They split the old casing, but managed to replace it.) The total cost of the re-entry was $86,000.

This is a man who knows what he is doing and looks after his investors' interests.

Chapter 5

PRODUCING PROPERTIES

Every day big money is made in the sale of producing oil and gas properties. Some people selling these properties and their overriding royalty interests were, once upon a time, shoe salesmen, real estate agents, cosmetic or drug salesmen, etc. Their game is hustling for the enormous fast buck. One thing for sure — most of these sales persons know nothing about the oil and gas business. They do know about the art of "selling."

A Common Scam

One angle of these exploitation sales (I would go so far to say, the most frequently employed) is the promoter finding someone or some company wishing to sell an oil and gas producing property; let's assume, as an example, a company wants to sell a producing property at an asking price of $900,000. The promoter tells the property seller: "I will try to sell it for you for a 2 percent sales commission, which will be $18,000." The seller agrees.

The promoter casts his eye about and locates Investor One, who has some money, but not enough to buy the whole property at the seller's asking price. I should mention here that the promoter has told Investor One he is selling a producing property for $1,050,000.

Now, the promoter tells Investor One, "If you can find one other investor to join you for this buy-out I will give you $50,000 for helping me sell it." (The promoter tells the

DRILLING PROSPECTS

investor that he is to receive 7 percent sales commission, i.e., on the $1,050,000 property sales, a commission of $73,500.)

Naturally, giving this information to the investor the promoter says, "You will make more money than I will make. You will receive $50,000, and I, the promoter, will receive only $23,000."

The investor, seeing he can make $50,000, calls upon a friend who has money. Investor One and Investor Two agree to give the purchase a try.

The promoter is happy, and the investors are happy to acquire a producing property. The sales promoter, to profit $118,000, sets the procedures for the happy investors (it could surely happen this way!):

1) The promoter has offered Investor One $50,000 cash, tax free (the investor is to receive his own money).

2) The promoter will arrange to have a legal sales agreement signed by the property owner, the oil company, which states the selling price is $900,000 minus 2 percent sales commission.

3) Then, the promoter files an "assumed name affidavit" with the county (for instance, The Box Oil Company) and is set up now to do his business.

4) The promoter arranges the transaction through a bank trust officer — because it will appear more professional to the buyers and the seller.

5) The promoter approaches the seller and states he has a firm commitment for the $900,000 sale of the producing property.

6) The promoter has the purchasers (the investors) sign the sales contract with the Box Oil Company which he hands to the trust officer of the bank to hold.

7) Then the promoter draws a sales contract between the Box Oil Company (the "buyer" of the property) and the seller (the oil company).

8) The promoter calls Investor One and tells him the deal is set to close, the sales contract between the Box Oil Company and the investors, as buyers, is signed; the property sale is listed as $1,050,000. An affidavit states that

Producing Properties

a check, or checks (from one or both investors) must be received by the trust officer of the bank, to total the amount of $1,050,000.

9) The trust officer holds a check of Box Oil Company in the amount of $50,000 payable to Investor One.

10) When the trust officer receives the checks from the investors, statisfying the agreements with the Box Oil Company, the promoter will receive his money.

11) The promoter draws up a check on the Box Oil Company account, payable to the original seller, the oil company, in the amount of $882,000 (the $900,000 asking price, minus 2 percent — $18,000 — sales commission).

12) The legal transfers of the properties from the oil company through the bogus company, the Box Oil Company, takes place, and the producing properties are transferred ("sold") to the investors. With the property in the hands of the investors, the promoter may now withdraw from Box Oil Company accounts for his use, and for deposit in his own personal account (or under his mattress, or in the wall safe of his palatial home) amounts equal to $18,000 plus $100,000. In other mathematical terms:

To Box Oil: Company	$1,050,000	Investors' purchase money
From Box to Investor	- 50,000	bonus for securing Investor Two
From Box to Oil Company	- 882,000	sales price of $900,000 for producing property, less 2 percent commission
Total for promoter	$ 118,000	

How can the investor be protected from becoming entangled in this kind of transaction? First and foremost, good sound business judgement must be employed. The intuitive, aware investor will ask to see a selling contract agreement between the promoter and the oil company, or

DRILLING PROSPECTS

the original seller. If that transaction looks favorable on paper, the next step is to request run tickets from the sale of oil. As for gas wells, the aware investor asks for copies of monthly gas-purchase statements. When these steps are completed, the investor is in direct contact with the producing property that is up for sale.

Check the Run Tickets

Some promoters will show one month of the oil run tickets. This is not acceptable to the aware investor. Such an investor asks for at least the past six months of these tickets, even a year.

Here's another example. The promoter is trying to sell four wells that once were good wells. Each well had three 400-barrel storage facilities — 1,200 barrels of oil could be stored at each well site. An investor was told that these wells were producing 70 to 80 barrels of oil a day. The investor is shown the oil run tickets for the month of April and sees that 2,160 barrels of oil were sold from each well. For the 30 day period, this averages 72 barrels of oil (2,160 barrels at $30 each totals $64,800). However, when the investor attempted to obtain the oil run tickets for the month of March, there were none — no oil was sold during that month.

Here is what happened. Oil was sold at a rapid pace in April. Actually two months' oil production was sold in one month, which means the wells were producing only about 36 barrels of oil per day — averaging the one-month sales over two months — and for a thirty-day period (April) the production would in reality be 1,080 barrels. At $30 each, the 1,080 barrels a month is $32,400, or half of the $64,800 shown to the potential investor. Promoters often try to pull this kind of scam on an investor.

Let's take one-year's production in dollars and cents, using the April run tickets as a guideline.

Four wells, using the same production sales, equals $259,200. For the twelve-month period the promoter could

Producing Properties

Modern production rig arranged for workover service.

DRILLING PROSPECTS

have shown the investor a remarkable total oil production sale of $3,110,400. Because the wells were actually producing half of the total shown this must be cut in half. The investor receives $1,555,200. Therefore, check the monthly run tickets against the daily gauge reports that are listed by the well pumper in the field. This gives a true indication as to the amount of oil sold versus the amount of salt water produced. Although the salt water must be disposed of, it is an expensive process and can easily cause operation costs to be too high to be profitable.

Production facilities must be inspected and the true findings reported to the investor. The oil storage or salt water tanks may leak, or there may be leaks in the oil and water lines. A visual inspection will help the investor realize the true condition of the equipment. Also, such inspection will show the investor the work that needs to be done at the site.

It's not a bad idea to inquire how long the production facilities can operate in their present condition.

Workover

For those unfamiliar with gas well/oil well terms, a workover* is actually, physically reworking the well. In this process, a workover or completion rig is used. This operation can become very costly, and must be checked by the investor. If the workover operation is too costly, it may be one of the reasons the lease is for sale. On the other hand, such a well purchased at a low price and then reworked in the proper manner can end up being profitable. Every day countless numbers of people look for new ways to complete a well, finding something that someone has overlooked.

I know one man who has a well that he's dubbed his million dollar well. He tried to recomplete a well for another man and was not making any progress. The recompletion man asked the owner about coming up the hole to shoot a small two-foot zone, or strata. The owner nixed the idea and favored plugging the well. The well owner sold the

Producing Properties

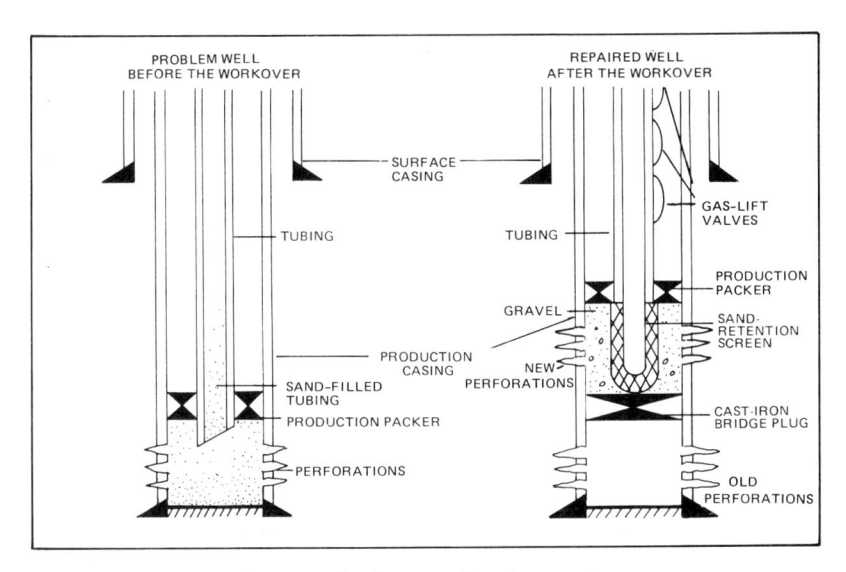

Before and after well schematics.

recompletion worker the well; the owner agreed, literally almost giving the well away, just to be rid of it. The new owner came up the hole, shot the two-foot zone and made a fairly good gas well. It has been producing for nearly ten years now. If you look at every aspect of producing properties before buying, you may find a million-dollar well — cheap.

Liens on Producing Properties

Another item for consideration, if you are interested in buying producing properties, is to determine if there is a lien on the property. One needs only to question the seller and request documents stating that no lien exists. If there is one, what is its amount?

It should be determined that there is no suit against the producing property or a proposed suit pending. (A proposed suit pending is called a *lis pendus*.) Naturally, this could very well put a cloud on the title. A valid outstanding claim

DRILLING PROSPECTS

or encumbrance would affect or impair the owner's title. Such encumbrance may even prove the property worthless.

Liens on a producing or non-producing well may be filed for any number of reasons. The most common reason is failure of the operator to pay for pipe or equipment. Very often a small company or operator will start out paying bills with no problem, but find after a time that money is short and they get behind.

I am acquainted with one company that drilled three wells, one right after the other, paying the drilling contractor for the first well, but then ran short of money and was unable to pay him for the other two. He filed a lien on the wells.

Get All the Information

I must explain another incident that would create great losses to the potential investor. A major oil and gas company lost over seven million dollars in such an incident, not including their time and legwork. However, they did not spend sufficient time investigating the wells. (Always be thorough!) If a major company can get trapped by not investigating every item, the small investor is even easier prey.

An independent oil operator had several wells located in South Texas, producing gas at a very good rate for a few years. The independent decided to sell the wells and approached first a major oil and gas company producing wells within a ten-mile area of his operation. Of course, the company was interested and asked the independent to bring in the well logs, mineral leases, and production records to their Houston office so that a complete study and report could be made. As instructed, the independent took all of the data he had on the wells to the company's office — *with the exception of one set of invoices.*

The engineers immediately studied the material provided, and the production reports checked and matched perfectly. All of the wells reported production out of one

Producing Properties

zone, and that is what the Railroad Commission reports indicated. Even though these wells had produced for several years, the independent had a lucrative selling point. The well logs showed four other potentially productive gas zones in each well. There were records filed with the Railroad Commission indicating no production had been taken from the gas zones. Everything looked good to the oil and gas company executives. The deal was closed, the independent oil operator was given payment in excess of seven million dollars, and the operation was taken over by the major oil company. After a couple of months passed, all the paperwork had been processed, and the oil company decided that production must be increased to gain some of their investment back.

An engineer of this major company, from its office in Corpus Christi, was informed of the work procedures. It was time to begin perforating some of the four potential gas zones that had looked so good on the well logs. The well logs were sent to the engineer in Corpus Christi. He was authorized to initiate the work by perforating the first well and testing the zone. The testing results would be sent to the company's Houston office before proceeding with more work.

This company would never have lost those millions of dollars had the first engineers' pre-buy study been proper and faultless. They thought they had found a bird's nest on the ground, and, making a good show in their report, would advance up the company ladder. Obviously, the ladder broke.

The Corpus Christi engineer had all his equipment lined up on the first well and proceeded with the work. He had the well log in the perforating truck, and they ran the perforating gun down the hole. When the gun started into the first potential gas production zone, the truck began receiving a beeping sound from the collar locator, an indication there was an indentation in the pipe. The engineer did not believe this a possibility, so he told the perforating truck operator to start his graph machine and run through the zone again. All indications showed that the

DRILLING PROSPECTS

zone had already been perforated. So they moved to the next zone and the identical situation occurred. They ran the gun with the collar locator through all four potential gas zones. They had all been perforated. The operation folded.

The engineer reported to the Houston office and explained what he had found. He showed a picture of what was down the hole. The following morning, a company plane rushed to the Corpus Christi area with engineers and vice presidents of the firm. They examined the sad evidence. All the zones in this one well had been perforated. It was then decided to run just a collar locator in the rest of the newly purchased well. Of course, this cost additional money, but the final results were essential in verifying all the potential zones had indeed been perforated in every well. To make matters worse, the independent oil operator could not be found. In fact, his whereabouts have still not been discovered.

He got away with seven million dollars.

You will remember, the independent operator gave the major oil and gas company everything *except* one set of invoices. There were invoices for the perforating charges on each well and each production zone. These were later tracked down by the oil company. The independent never filed the mandatory recompletion forms with the Railroad Commission in the State of Texas. All the Commission had on file was a record of only one production zone perforated and producing.

The engineers doing the study for this sale made two grave errors. One very good indication that something was amiss was the production history. As I have stated before, 98 percent of the time, production will have a steady decline. These particular wells had a steady decline for about two-and-a-half years. Then the production rate picked up. In fact, the wells gained too much in production. Any irregularity of this sort suggests that the investor should investigate further. The second mistake the engineers made was taking a person's word as to the condition of the wells' downhole. On a deal of this size, it seems mandatory a clause should be in the sales agreement which would

Producing Properties

deposit the money in trust, or escrow, until all aspects of potential future production could be checked out. The oil company could have run a collar locator down all of the wells much cheaper!

Several essential tools and pieces of equipment assist the investigating process in the purchase of oil or gas properties.

Perforating Procedures

After the drilling of a well has been completed and open hole logs have been run, a decision must be made. If the well is to be plugged, the procedure must be carried out as required by government agencies. If the well is to be completed, then casing must be run in the hole for production purposes. The casing or pipe is run in the hole carefully, and an accurate tally of the casing is recorded. The reason the tally must be perfect is obvious. If the hole is drilled to a total depth of 10,000 feet and only 9,600 feet of casing can be put in the ground, something is wrong (part of the hole has filled up, or caved in). Ninety-nine percent of the time each joint (piece) of casing is a different length and joined together with a collar. The collars will furnish a reference point when a cased hole log is run, or a potential production zone is about to be perforated. After the casing has been run in the well and cemented, most often a cement bond log is run on the well, and a collar log also. The cement bond log indicates if the cement is where it should be and has a good bond to the casing. The collar log indicates the location of each collar in the ground and may be correlated with the open hole logs as to the position of certain collars in relation to the exact depth of the production zone to be perforated. This correlation procedure is used for precise judgement in the perforation of a zone, locating the exact spot for the perforating gun to penetrate the proposed zone in the proper place.

The collar locator is a device that may be run in the hole either by itself or on the top of some other tool, such as a

DRILLING PROSPECTS

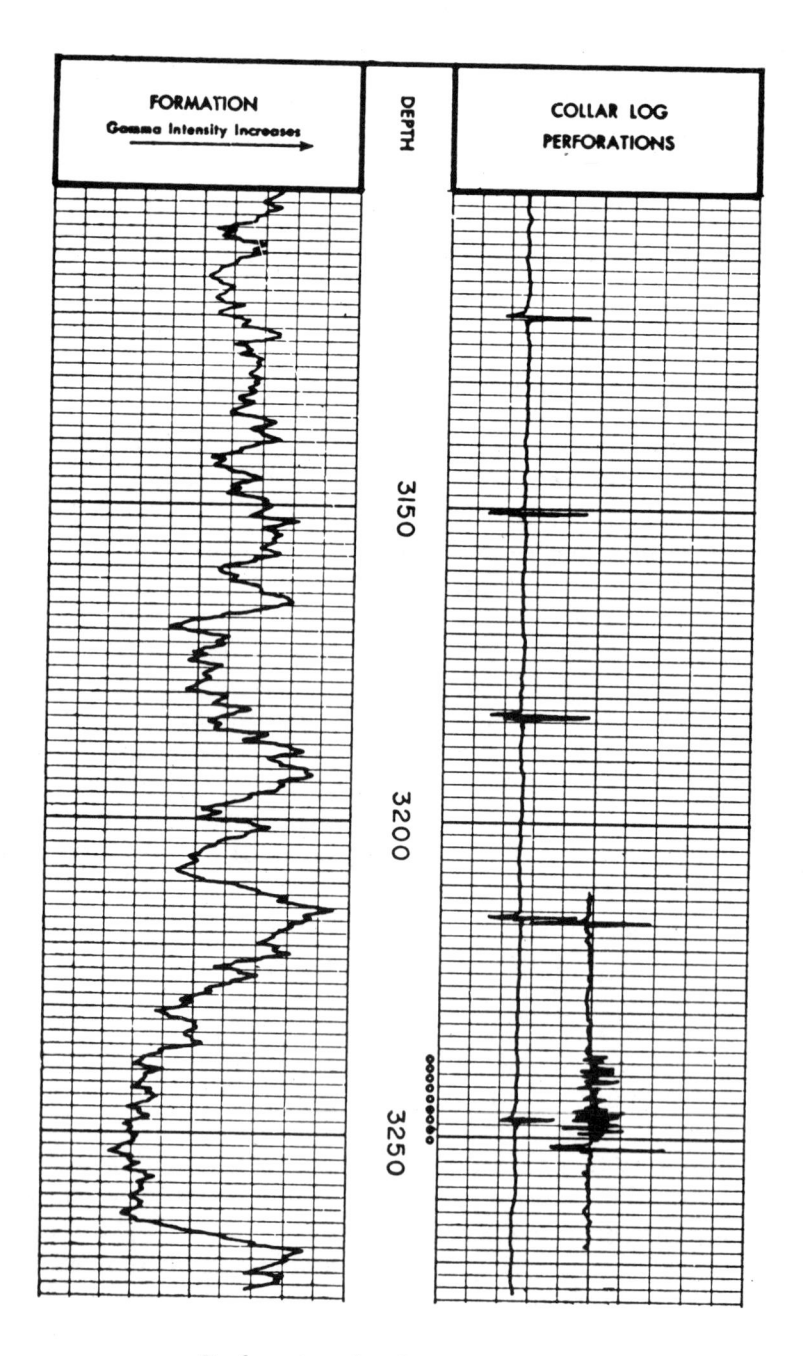

Perforation depth control log.

—78—

Producing Properties

perforating gun. It is designed in such a manner as to pick up indentation in the casing, or the casing connection (collars). As the locator passes a collar, a bell will sound and a mark will be made on the logging graph by a pen. The graph also records the depth at which this procedure is taking place.

I was taught a certain procedure in well perforating that I think is important. Most young engineers today do not know what I am talking about, or why. It is very simple and furnishes a record. After perforating, pull the gun up about 100 feet or so and wait about 15 minutes. Let things settle out in the hole after firing the charges. Then start the recorder in the perforating truck and go on down the hole to the spot that was perforated. As the collar locator goes by the zone that had been perforated, if the gun fired properly, the bell will sound and the indentation will be inscribed on the graph. One might ask, why do this? The answer is simple. You will have a picture of the exact depth of the perforations and gain the knowledge if the gun fired or not. The picture produced by the collar locator may then be correlated with the open hole logs for proof positive as to the exact spot the perforations were made in the casing and ground.

Let us examine the reasoning. A collar locator may be run in a well under pressure and not bother a thing. It's just a small expense and will indicate perforations and their locations.

Part III

PROTECT YOUR INVESTMENT

Chapter 6

COMPANY EXPENDITURES AND YOUR MONEY

It is rarely advisable for an investor to hand over his money to a promoter or oil company, then sit back and trust that it is being handled as economically as he would do himself. If the investment is important to the investor (and we must assume that it is), I recommend that he maintain an interest and a curiosity even if he is receiving healthy checks in the mail.

He should ask himself if he is receiving as much as he should be. There are ways that he can be sure.

The A.F.E.

The Authorization for Expenditures (or A.F.E., as it is referred to) defines the monies spent for a drilling prospectus. Investors should be aware of its pitfalls. There are a number of common abuses involving the A.F.E. From time to time, the A.F.E. also may simply get out of hand within the best of organizations.

There are numerous types of A.F.E. forms used in the oil and gas business. However, they should all reflect the expenses pertaining to the drilling venture. I have included a sample of a typical A.F.E. on pages 84 — 85. In order to familiarize yourself with the items on the A.F.E., study it carefully. Look for any item listed where overcharges may be difficult to uncover.

Also examine it with a questioning eye to determine if

PROTECT YOUR INVESTMENT

A.F.E. — Authorization for Expenditure

GOPHER
Energy Operating Company

Well Name & No. _BURP #1_ AFE No. _111-_
Location _Ohio, Texas_ Date of Request _MAY 1983_
3600 feet - 40 acres Estimated date of completion _Oct. 1983_

Feat. No.	TANGIBLE EQUIPMENT (A/C 312 & 313)	Amount Dry Hole	Completion
001	Conductor/Casing		
002	Liner		
005	Intermediate Casing		
006	Production Casing _3600 Ft. 4½ - 10.5# - 3.10/ft._		_11,160_
007	Surface Casing _282 Ft. 8 5/8 - 24# - 5.30/ft_	_1,500_	
101	Bradenhead	_300_	
105	Casinghead - Intermediate		
106	Casinghead - Production		
110	Wellhead		_1,500_
120	Surface Safety Equipment		
201	Tubing _(USED) 2 3/8 - J-55 3600 Ft. - 1.00 Ft_		_3,600_
202	Tubing Equipment		
210	Packer		_1,300_
301	Sucker rods _(USED) 3/4 - 3600 ft. .50/ft_		_1,800_
305	Sub-Surface Pump		_1,500_
401	Flowlines and Connections		_1,500_
405	Separators		
407	Treaters		
409	Storage Tanks _2 - 202 TANKS_		_6,000_ / _4,075_
415	Free-Water Knock-Outs		
418	Dehydrators		
420	Intermitters		
425	Buildings		
430	Gas Metering Equipment		_3,000_
435	Gathering Lines and Fittings		_3,000_
501	Pumping Units		
503	Engines		_15,500_
504	Pumps		_2,500_
505	Compressors		
	Total Tangible Equipment		
	INTANGIBLE DEVELOPMENT COSTS (A/C 910)		
001	Legal Fees	_1,000_	
002	Geological Fees	_1,000_	
003	Engineering Fees	_1,000_	_1,500_
004	Engineering Fees - Other		
005	Other Professional Fees		
101	Trucking and Hauling	_1,000_	_1,000_
102	Supervision and Overhead		
103	Supervision and Overhead - Other		
104	Labor	_200_	_2,000_
201	Surveying	_200_	
202	Road and Location	_1,500_	
203	Pits	_1,000_	
204	Surface Damages	_1,500_	
205	Surface Leases _40 ACRES @ 30.00 ac_	_1,200_	
301	Turnkey Drilling	_28,800_	
302	Footage Drilling		
303	Daywork with Drill Pipe _12 hr. @ $150./hr_		_1,800_
304	Daywork without Drill Pipe		
305	Fuel		
306	Purchased Water		
307	Bits and Reamers		

Company Expenditures and Your Money

Authorization for Expenditure
(Page 2)

401	Drilling Mud and Chemicals		4,000
501	Drill Stem Tests		5,000
502	Mud Logger	2,200	
503	Logging - Open Hole	3,200	
504	Logging - Cased Hole		1,200
505	Coring and Core Analysis		
601	Pumping Services *Casing Crew*		900
602	Pumping - Cement Conductor		
603	Pumping - Cement Surface Casing	1,650	
604	Pumping - Cement Intermediate Casing		
605	Pumping - Cement Production Casing		4,000
606	Pumping - Cement Liner		
610	Cement - Conductor		
611	Cement - Surface Casing		
612	Cement - Intermediate Casing		
612	Cement - Production Casing		
614	Cement - Liner		
615	Float Equipment		280
16	Scratchers and Centralizers		400
701	Completion Unit *6 days @ 1,000 a day*		6,000
702	Completion Unit		
703	Acidizing		6,000
704	Fracturing		
705	Fracturing Fluids		
706	Squeezing		
707	Perforating		1,200
710	Rental Equipment - Tools		1,500
711	Rental Equipment		800
712	Rental Equipment - RCC		
720	Load Oil		
725	Supplies		
730	Abandonment		4,000
801	General and Administrative	200	
802	Insurance	1,100	
803	Interest		
810	Reconciliation of AFE Advances		
	Total Intangible Development		
	Turnkey Contract Costs		
	TOTAL AFE COSTS		

The above is an estimate only, and it is understood and agreed that this AFE shall be your authority to drill said well and incur such expenses as are necessary, whether the cost is more or less than the above sum.

ACCEPTED AND AGREED TO:

EXECUTED BY	WORKING INTEREST
	DATE

—85—

PROTECT YOUR INVESTMENT

there are discrepancies that suggest you are not getting the full benefit of your investment. The A.F.E. can be misleading. I'll give you an example, using the hypothetical mineral lease I've used before, the lease of 500 acres.

The oil company intended to sell the lease at $150 an acre, a total of $75,000, but selling one well at a time. It seems that there was some logic to their scheme. The promoter gave the investor the prospectus to look over, with the A.F.E. to examine, as well. The promoter is offering one well on 160 acres. The lease cost is $75,000; he did not lie. The deal is that he's selling an interest in one well on 160 acres out of a total of 500 acres. The actual lease cost for the 160 acres he is attempting to sell the investor is therefore, $24,000. If a sharp investor does not catch this bit of fast-talk, this promoter may sell the same 500 acres three times and possibly clear around $145,000. More than pocket change for a few days work.

But this is not the only way the A.F.E. can be deceptive. Shrewd promoters are always finding new angles for various points. For example, consider the location work, which involves building roads and preparing the location for the rig. Examine the amount allowed for this carefully. The cost of the work depends on the size of the rig to be used to drill the hole, how deep the hole is to be drilled, and the roughness of the terrain at the location site.

You may recall we have referred to a 300-acre mineral lease that didn't appear too profitable. The engineer had the dirt contractor buy an interest in the well using "overpayment" as incentive. There is the problem of kickbacks when we keep in mind that most contractors will buy into any job they can that gains for them another incentive. Hand $1,000 or so to the person in charge of preparing the location for the rig, then jack up the bid price of the work, say $2,000. The dirt contractor is going to make a little extra money because he has his man hooked.

I was employed by an oil company and observed a location built near Kurten, Texas. The location was on dry land and cost over $130,000. The correct average cost for this location should have been $34,000, to a top of $40,000.

In building a road and a location for a 15,000-foot drilling rig, rocks are placed on the road and location to support the heavy equipment. The road to this location was less than one-fourth of a mile long, and there had been no trees to clear for the roadway. I measured the road and the location. Rocks would have to have been buried eight feet in the ground to have necessitated a cost of $130,000 for the location. This was not the case; it was just a standard location job.

I reported my findings — that the price was at least $90,000 too high.

Companies Furnishing Their Own Rigs

Another situation that does not impress me is an oil company that sells drilling ventures and furnishes its own rigs. At one time, I thought it an ideal set-up, but I have since changed my mind. On the A.F.E. there is a rig price of $8,200 per day, plus fuel. What I found was that the same rig could be rented from an outside contractor for $7,200 daily, plus fuel. If an oil company owns six drilling rigs and they are charging this higher price, that's $6,000 they stick in their pockets each day. In 340 days, the oil company has made an unbelieveable $2,040,000 bundle from their investors in overcharges.

There are still other disadvantages to consider. Moving a rig is an expense most people take for granted. For example, say we are to drill a well in South Texas. There are rigs working in the area, and we luckily locate one that will be available to fit our drilling needs. It has to be moved 44 miles, at a cost of $22,000. As it turns out, we can't use that rig because one of our own company rigs will be available. However, our drilling rig is 208 miles from our location; but we have to use it. Moving the rig costs the investors $58,000 and if each rig makes four unnecessary long moves then the company has cost the investors $864,000 too much money.

PROTECT YOUR INVESTMENT

Overcharging for Pipe

The pipe on an A.F.E. is another item that merits concern along with the wellhead (christmas tree) cost. Let's assume we are going to drill a well to a total depth of 10,000 feet. We log the well, and it looks like a production zone has been located. We set 10,000 feet of 5 1/2-inch casing on the A.F.E. shown to the investor. The price per foot for the 5 1/2-inch casing was $7.95. On investigation, I find a low price of $5.65 per foot for the same grade pipe listed on the A.F.E. That is $2.30 difference and would total an overcharge of $23,000 to the investors.

Why does this happen? Perhaps a member of the board of directors of the oil company owns an interest in a pipe supply company. The investor has been hooked again. This example could also apply to the wellhead and tubing contracts. If we inquire to see that the prices given us are the best the promoter can get, we can encourage the investor's confidence in the venture.

Coring

Another item included on the A.F.E. is *coring.** This process assists the geologist and the engineer to evaluate the potential production zones in the best possible way. As an example of coring procedure, we drill a hole with a 7 7/8-inch drill bit. The geologist calculates an encounter with a possible production zone at a 7,600-foot depth. The rig driller and the tool-pusher figure they have drilled to a depth of 7,595 feet, so drilling halts. They pull up the hole and circulate the drilling mud until all drill bit cuttings have been cleaned out of the hole. The driller then pulls all the pipe out of the the hole and removes the bit. Then, the core lab technician attaches a core barrel* to the bottom of the drill pipe where once the regular bit was attached. A barrel may be of different lengths, but in this example, we will attach 40 feet of core barrel. These barrels are of different diameters, however; we drill with a 7 7/8-inch bit

Company Expenditures and Your Money

so we will run a 3-inch diameter core barrel. The hollow barrel will usually have a diamond cutting-head on the bottom in order to cut the sample, and also a catcher. The catcher is used to retain the sample inside the core barrel when bringing it up to ground level.

The driller lowers the core barrel into the hole and cores 40 feet, to a total depth of 7,635 feet. Then he pulls the drill pipe with the barrel. The core lab technician takes the barrel from the pipe and, having removed the 40 feet of sample, goes to the lab to prepare his findings. The driller puts his 7 7/8-inch bit back on and continues with the drilling operation.

What has taken place is that a 40-foot section of the proposed production zone was cut away, brought to the surface in one piece, and sent to the lab for analysis. The results tell the geologist the composition of the section, the earth's makeup, the location of gas and oil, their content percentages, and water content. The geologist may be able to advise the engineer how the zone is to be treated to get the products out of the ground. The core data received will also be compared with the well open hole logs. This is one of the best procedures used for finding oil and gas and bringing it from the ground.

There's a catch, though!

I obtained an old A.F.E. to examine, as another example of investors' bad luck on being taken into schemes. This A.F.E. states: *coring, $30,000.* It does not state how many feet were cored or if more than one zone was to be cored. Although that was a hitch, it wasn't the problem. The main problem was that there was no coring done at all. So the investors were hooked for $30,000.

You, as an investor, need to know what is happening. How is your money being used? Are you getting what you paid for? If the promoter has coring on the A.F.E. at "X" amount of dollars, demand a copy of the core results, or threaten to pay nothing. The promoter might tell you that you wouldn't know what you were looking at, but it is yours to see, you paid for it. If you don't understand an A.F.E., find someone who does.

PROTECT YOUR INVESTMENT

Consultation

The A.F.E. will have an item that could be listed as (a) engineering supervision, (b) rig supervision, (c) daily supervision, or (d) technical supervision. Request if a "consultant" is to be employed in the drilling and completing of the well. (Although there are a few good consulting firms in the oil and gas business, as far as I'm concerned this *consulting profession* has become a joke.

Perhaps I should explain why I do not believe that consultants are necessarily an asset.

I worked with Millican Oil Company as a field superintendent. I had my own district, and there were two other districts, with men like myself in charge. We worked well together. If one district man needed help, he knew he had co-workers he could count on. (I might add, the two superintendents assisted me more than I had the opportunity to aid them.)

At one time, I had two drilling rigs working, two workover or completion rigs operating, and a crew laying pipeline. I had a personal interest in this oil company, and there was one very important factor in this whole operation. I was required to report by telephone each morning to the president or vice president of the firm. If, for some reason, I could not reach one of them, I had private telephone numbers of the investors whom I should contact. Millican Oil Company reported to their investors each day. I knew one man who took care of outside investments for a sporting goods firm. I would call him each Sunday morning, and he would have his report ready each Monday morning. Millican Oil started out with nothing and made money along with their investors. We watched all of our pennies and worked hard. A consultant, on the other hand, has no company interest, nor does he have investor interest. He is merely on a job, and I think receives too high a payment for the work he does, which is primarily taking orders.

A drillling rig has a tool-pusher; and a completion rig, or workover rig, has a pusher. If the oil company has competent field people, a third-party watcher is

Company Expenditures and Your Money

unnecessary. Consultants may charge an outrageous fee, approximately $350 to $400 a day, plus expenses. In a 20-day period, at $350 per day plus expenses, this could add as much as $8,000 to $9,000 added cost to a venture. What's worse, I have observed consultants taking cash kickbacks for using a certain company's frac tanks. I have also seen consultants, in the oil drilling business for two or three years, supposedly running a job., i.e. supervising the work.

To perform any technical job in the most proficient manner, qualified people are needed. If using a consulting firm, references should be checked carefully.

Gas Pipelines

At present, natural gas is difficult to sell, and the price of oil is down. However, I predict it will not stay this way. The glut will end, and the drilling rigs that are stacked will be placed back into operation. The oil and gas business is profitable, and more so if people invest their money wisely.

Let's say you have invested in the one well on the 500-acre lease described previously. A gas well has been completed and you are ready to sell the gas. A pipeline company is located, and a contract is negotiated and signed with it to remove a specified amount of gas each day at a determined price. But there is a slight problem. The investors who own the well have to lay more than two miles of pipeline to get the gas into the pipeline carrier. That's not a major obstacle, since the well is good and there is a market for the gas. However, with more money from the investors being used, the pipeline is laid. Finally the sale of the gas has begun, and the investors are getting returns from their investments in the one well.

I have said earlier that there is logic in the promoter selling one well at a time. Here is another hooker for the investors of the first well. The investors in the original well paid for pipeline right-of-way plus the pipeline. While the promoter was in the process of laying the line, they installed on this line two to four extra valves above ground.

PROTECT YOUR INVESTMENT

The investors do not know these valves are installed. They just know the amount of money they are charged for the installation of the pipeline.

I've also stated the investor should ask for an option in the second or third well. This will help stop what could possibly take place — call it preventive measures. The promoter will sell the second and the third drilling venture on the 500-acre mineral lease. To the investors on the second and the third drilling venture, the promoter will use as a selling point — you guessed it — the pipeline.

They will use this scheme in two ways. First of all, the promoter representing the oil company will tell the investors, "We already have a pipeline, and you can pay us the prorated share of the pipeline." Actually, what they do is sell something they don't even own. The promoter makes some extra money fast. What the promoter really does is to charge for transporting the gas in the pipeline; he is not selling the interest in the line. The pipeline fashion results in some big bucks that should go to the investors on the first well. However, they never see the money.

When selling the second and third drilling ventures, the promoter will tell the potential investors, "We already have a pipeline and charge 17 cents per thousand cubic feet to transport the gas." The two ventures are completed and selling 600,000 cubic feet of gas a day. The promoter is receiving 600 × 17 cents off each well a day for transporting the gas through someone's pipeline. They are taking in $204 a day unbeknownst to the investors on the first well.

But wait. There's more. I stated that the promoter could have put in three of four valves. Let's say a company on the adjoining lease produces a good gas well. It needs to get its gas to the pipeline that is purchasing the gas. Now the promoter charges this second company the same price to haul its gas. It we assume this gas well will also produce at the rate of 600,000 cubic feet a day, the promoter has another $102 a day coming in (a total of $306 a day going to someone other than the investors). A 1 percent interest in the more than two-mile pipeline could have returned $1,710. If anything goes wrong on this lengthy pipeline, such as

—92—

leakage, repairs, or whatever, the investors on the first well will more than likely pay all of the bills. Did the investors ask questions? Did the investors get the documented facts? Did the investors make money? No.

Hauling Oil

Oil from a producing well is handled in an entirely different way from gas. If various investors are in each producing oil well, each well must have its own oil storage facilities. They cannot be combined, due to various funding and money disbursements. Before oil is sold or loaded in the hauling truck, a shake out is made on the oil in the storage tank. This process (shake out) is used to determine how much foreign matter is in the oil. The foreign matter is termed Basic Sediment And Water (B.S.&W.) Most often the percentage of foreign matter is low (one or two percent is normal).

An example of what might happen: The hauling truck will normally carry 180 to 190 barrels of oil. (The shake out process indicates 2 percent B.S.&W.) The hauling ticket made out by the truck driver has the B.S.&W. recorded at 9 percent. His ticket indicates he loaded 180 barrels of oil. This causes 7 percent of the 180 barrels that the oil company (or the investors) will not get paid for (7 percent × 180 = 12.6 barrels) — 12.6 barrels × $30 each equals $378. This slight discrepancy may go unnoticed if it happens just once a week. But in a year, with this shortage occurring once a week, the loss would be valued at $19,656. If an investor had money in six wells and this situation existed at each of the six wells, the loss adds up to $117,936 a year. I think this would more likely occur where a bonus is paid for the oil.

An investor can and should ask for an audit of the run-tickets versus the daily tank gauge reports. This will give an indication of potential losses if a pattern of oil shortages exists. For that matter, any shortages should be investigated.

Chapter 7
INCOMPETENCE

The investor is always led to believe that he is dealing with professionals. The promoter suggests an investor buy an interest in the drilling venture because he knows what he's doing. "We're the best," he will say. However, incompetence has cost investors hundreds and thousands of dollars, and possibly even millions of dollars. And it occurs with far greater frequency than one would expect.

Contractor's Errors

I can recall a number of notable disasters from my own experience. Within one year, not one but two drilling locations were constructed on the wrong piece of land, complete with roads built to the locations. Both mistakes were authorized by the same engineer. The most expensive of the two actions had to be remedied immediately. The well site had to be cleared, damages had to be paid to the landowner, a fence had to be repaired, and a new rigging location built at the correct location. This inexcusable error cost over $100,000 and made an enemy of a landowner. Luckily, the second wrong-location construction was a much less expensive blunder.

Employed as field operations supervisor for this company, I was in the process of checking locations that were supposedly ready to move drilling rigs onto. On this occasion, I could not find a particular location, and I called

PROTECT YOUR INVESTMENT

the dirt contractor. He informed me that the location was prepared and the water well had been drilled. I told him that I had personally visited the proposed well site, and I could find no sign that work of any kind had begun. The contractor met me, and we drove together to the location he had constructed. There it was, on the wrong property. (This was the second operation authorized by the same engineer who had fouled up earlier.)

I faced the problem of the landowner whose property had been messed up. I asked him what I could do to get this straightened out. Fortunately, he informed me it would be all right for us to leave everything as it was. He could use the pits and the water well for a water reservoir for his cattle, and he would use the drilling rig pad as a cattle feeding area in the wintertime. He was also glad to have a good, new access road.

Another dirt contractor was hired to construct the drilling location on the correct tract of land. This small mistake cost only $48,000 for the location and $3,600 for the water well, a total of $51,600.

Easements

Heading the list of serious incompetent mistakes is trespassing. I'll give an example. Assume a 500-acre mineral lease is entirely surrounded by land owned by people other than the lessors. But the 500-acre lessors have a right-of-way or easement to get onto their property using an old road. However, this old road crosses two ditches with old bridges that are unable to support the heavy loads that have to be transported to the drill site. The person in charge of road construction and the constructing of the location finds an alternate straight path to the proposed location, and the fact that it passes through two fences doesn't concern him.

The dirt contractor starts to work, cutting the two fences, installing cattle guards, knocking down trees, and

Incompetence

throwing up a road bed. He hauls in gravel to build up the road. An unknown man approaches the foreman, wanting to know what's going on. The foreman explains the operation. The foreman, in turn, is told to get off the man's property.

The promoter had failed to obtain a right-of-way or easement to pass through this property with an access road. All work on the drill site and the road ceases. The landowner has the promoter and the dirt contractor in a compromising situation. First, the right of way should have been secured. Damages have occurred, and will be paid for at a high cost, or the landowner will sue the contractor. A lawsuit is out of the question for the drilling operation — it would take too long to move through the courts. So the price is paid, and the work resumed. Another bungling error made by co-called "professionals."

The following is another mistake that should never have occurred. A well is drilled, considered a dry hole, and plugged. The dirt contractor has been watching the drilling progress of the well he constructed. After the rig is moved off this "dry" location, he brings in dirt equipment and begins work. His crew closes the mud pits and pushes mud all over the land, then cleans up around the location. However, the contractor has operated without any authorization. Some landowners want the drilling mud on their land, and some do not. If the mud pits are near a watershed area, the mud must be hauled out before the pits are closed in order to protect the marine life in nearby creeks and rivers. Should the contractor do any work on his own and violate any laws, or create land damage, it will all come back to the promoter. There is a chance that he may not have to pay for damages, but this is doubtful.

One of the most common mistakes made through incompetence is entering a tract of land before consulting all the owners involved, even if a good mineral lease has been secured. During recent years, I have witnessed many companies going onto properties without first contacting the persons who have an interest in the land.

PROTECT YOUR INVESTMENT

Let's assume that our 500-acre mineral lease had a leaseholder who did not own the land. The person owning the *surface* land has leased his rights to a rancher for cattle grazing. The rancher's lease contains the following provisions:
1) The entrance gate must be kept locked.
2) The roads on the land must be maintained.
3) The fences on and surrounding the land must be maintained.
4) All the ditches that feed water into a stock tank must be kept clean and open.

All of these provisions fit the rancher's needs.

The mineral leaseholder does not care what happens on (or to) the land. He merely wants a well drilled. The promoter has not been in contact with anyone concerned with the rancher's lease.

A dirt contractor is sent to construct an access road to the site and rig the drilling location. He cuts the lock on the gate, removes the gate, and installs a cattle-guard. The access road to the location crosses two ditches that feed water to the stock tank. The contractor fills them in. Now the rancher catches the dirt contractor engaged in this activity, who tells the rancher he is doing the work for the oil company. The rancher and the landowner together go to the oil company; they've got the company right where it will hurt, and they know it. The company is required to pay damages for trespassing and is required to undo the work, which will cost thousands of dollars. The promoter, who never bothered to secure proper rights, and the oil company have made two enemies.

But the trouble doesn't end here. Late one evening, after moving the drilling rig in, a truck tears away part of the fence, near the entrance. The driver does not report this accident to anyone, and the rancher's cattle stray from the pasture. The next day the rancher finds the damaged fence and calls the promoter. The lost cattle must now be paid for, and a new fence has to be built. This type of carelessness is unnecessary, but it occurs frequently.

Follow Through

As I have pointed out earlier, all kinds of problems can stem from mineral leases. Incompetence can enter into leases as well. For example, a promoter offers a mineral lease to the land department of an oil company; it is a drilling venture that looks promising. On paper, it looks like an excellent lease, with producing wells surrounding it. No one inquires why the mineral lease is suspiciously cheap. There's only one leaseholder to deal with; therefore, less paperwork. And the added attraction is the fact that there is 80 percent working interest for investors.

The promoter has made his own surface map, but he has omitted a few minor details. The land department takes up the lease, but it does not take the time to examine or survey the land or check around at the site. They rely on the maps the promoter brought into the office.

The first big mistake! This mineral lease is situated far out in the country, but it is property that a land developing company has divided into lots with houses constructed around a lake.

One problem exists, a major one. There is no available place to set up a rig to drill a well — except in the middle of the lake! Nowadays, certain laws require an oil and gas drilling rig to be situated a certain number of feet away from a residence, and a certain footage away from a gas line, and an established footage distance from an electric power line, etc.

The oil company cannot bring in a small sea drilling rig and set it up on the lake. That is out of the question. Other oil operators on the surrounding land will not allow a drilling rig to set up on their lease for a directional hole under the lake lease unless they are offered a large free interest in the well to be drilled.

The land department of the oil company, not having checked into the lease and not having actually examined the land, have proven themselves masters of costly incompetence.

PROTECT YOUR INVESTMENT

Incompetence in Well Stimulation

Investors should be aware of the procedures of well stimulation because they are what determine the life of the well. The illustration on the following page shows a typical well completion. It is intended as a reference for understanding the equipment on a completed well. It displays certain methods of equipment use.

In the fracing procedure of a well, one thing that is very important is the maximum desired pumping rate of the fluid and sand mixture into the formation of the production zone under relatively stable pressure. The pumping rate is designated as barrels per minute. As this fluid and sand is pumped down the hole and into the formation, the pressure that is required to inject this mixture into the formation is monitored at all times from several sources. Not only is the injection pressure recorded, but if they are fracing through the tubing, pressure is watched and recorded on the casing also. *This is very important.*

The pump rate, or volume per minute, has three controlling factors. First, the number and size of holes perforated through the casing. Second, the depth in the ground of the perforations. Third, the diameter of the pipe used for transporting the mixture to the perforation within the formation.

The desired volume to be pumped per minute sometimes will determine if the frac procedure is to be performed through the tubing or the casing. If enough volume per minute of the frac fluids and sand cannot be moved down the tubing at the predetermined required rate, the tubing must be pulled from the well. The frac procedure then will be performed down the casing. The stabilized pumping pressure will vary. However, it should stablize with each variance. The reason the pumping pressure varies (falls and rises) is because the volume of sand, usually started in the mixture at about a pound of sand per gallon of fluid, is increased to perhaps six or eight pounds of sand for each gallon of fluid; or higher if possible. All of the anticipated

Incompetence

TYPICAL FLOWING WELL COMPLETION

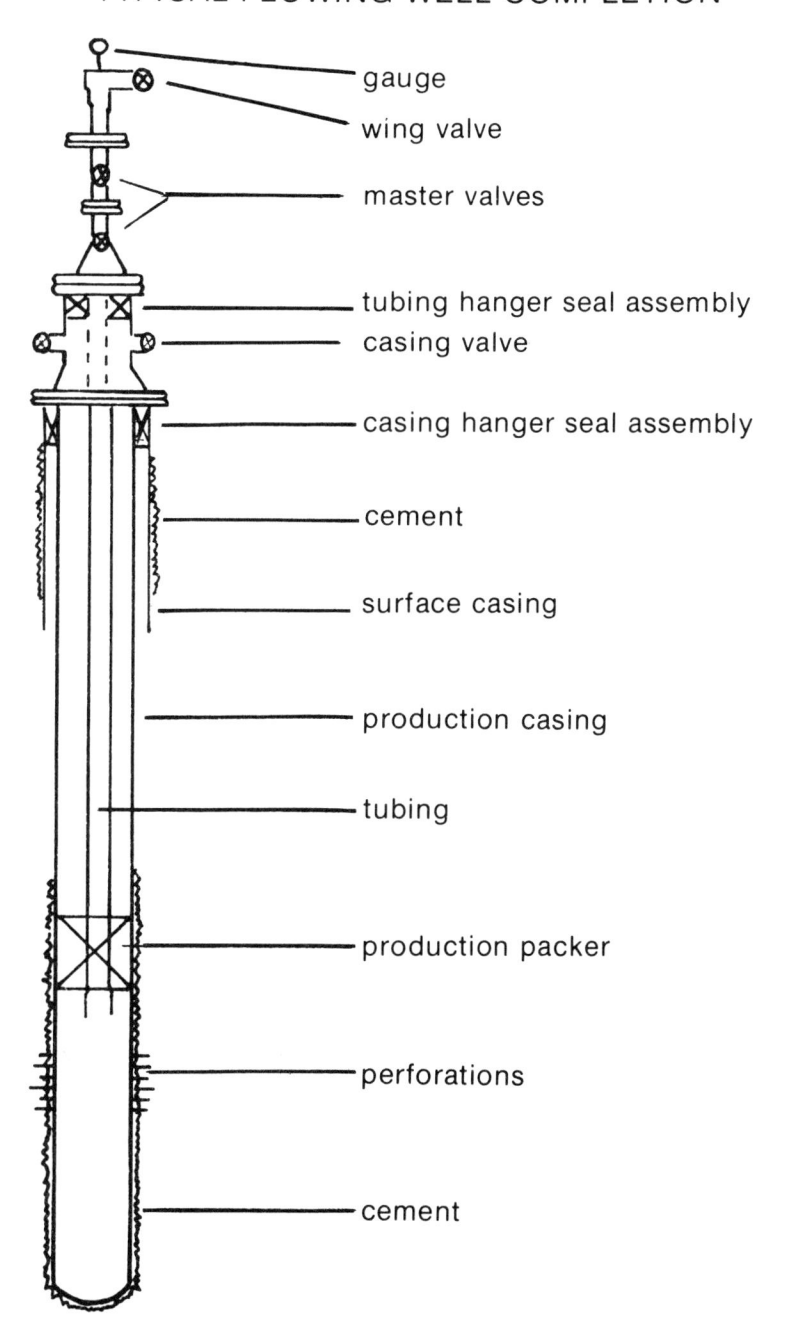

- gauge
- wing valve
- master valves
- tubing hanger seal assembly
- casing valve
- casing hanger seal assembly
- cement
- surface casing
- production casing
- tubing
- production packer
- perforations
- cement

PROTECT YOUR INVESTMENT

pressures to be encountered are, as closely as possible, predetermined by the service company.

Let us assume we are going to frac a well perforated at six-inch intervals, from 12,160 feet with 3/8 inch bullets. We will frac down two-inch tubing with a packer* set on the end of the tubing at 12,000 feet. The packer is a device, usually employing rubber, used to effect a seal between the tubing and casing, which also anchors down the tubing. Calculations have been made, and the maximum pressure we should obtain will be about 8,100 pounds while pumping the desired maximum volume. A pressure of 1,500 pounds will have to be maintained on the casing to hold the packer in its original set position. I cannot impress enough that every aspect of fracing is very important.

The main objective in stimulating a well through a frac procedure (or acid procedure) is to make the well produce oil or gas in an economical manner. Today a huge number of wells must have a frac job or acid job, or possibly both, in order to achieve an economical production rate. If gas or oil found in a production zone in abundant supply has a formation too tight to allow it to flow, a stimulation procedure is worked up with a service company engineer. Stimulation procedure will determine the life of the well in nine out of ten cases. All facts must be taken into consideration so as to produce a stimulation process that will obtain maximum results. The service company engineer needs to know the depth, the thickness, and the makeup of the production zone (content of limestone, sandstone, etc).

Also, it must be determined how many holes are perforated through the casing. Too, what are the sizes of holes: Also, will we frac down the tubing or the casing? Next, what are the tubing and casing sizes? And what pressures will they bear? What is the maximum pump rate (barrels per minute)? What are the maximum pounds of sand per gallon of mixture? And what grade is the sand? The final (and a most important) question is, what kind of wellhead equipment is on the well?

When all this information is collected, the engineer takes

Incompetence

himself to a laboratory. The laboratory sets up a well stimulation program that will be compatible for a specific well and its working equipment. A great deal of research and development work has created successful frac and acid procedures.

The service company engineer returns the lab's frac procedure program to the oil company. We will now have equipment worth millions of dollars moved to the well site. Frac tanks are trucked in and filled with water. Sand tanks (which do contain sand) are also trucked onto the well site. Then chemicals, pump trucks, blenders, and a monitoring van are all moved in. All lines for pumping procedure are hooked onto the wellhead and tested for leaks (calculated at higher pressures than those anticipated by the frac program). A consultant or oil company engineer sets up in the monitoring van. All recorders, radios, and earphones to all personnel are tested to function properly, and the digital read-out monitor is affirmed to be operating. A safety meeting is held. We're ready to frac the well.

Some frac procedures have cost a great deal of money, and some have lost thousands of dollars in profits.

One case involved a young experienced engineer. A service company was set to frac down the tubing and place 1,500 pounds of pressure on the casing side of the well, to help hold the packer set. They began pumping the frac job into the well, one pound of sand per gallon of fluid. The amount of sand was increased to 2 1/2 pounds per gallon of fluid. This is more than double the starting amount of one pound of sand. This could have started the problem they encountered.

In the final stage of this frac procedure, six pounds of sand per gallon of fluid was supposed to be obtained, with a maximum pressure of 8,000 pounds. But they never got that far. When they had attained four pounds of sand per gallon of mixture, the injection pressure jumped from 6,800 to 8,600 pounds, then fell back to 7,400 pounds. They pumped on for awhile. Finally, the engineer noticed his casing pressure had climbed from 1,500 pounds to 2,800 pounds. The frac procedure was halted and all lines were taken loose.

PROTECT YOUR INVESTMENT

The engineer tried to flow the well back. It wouldn't flow. The company vice president for operations asked me to examine the well site.

Examining the frac procedure, I discovered that sand was added at one pound per gallon of fluid. The next step, to raise the amount of sand to 1 3/4 pounds per gallon of fluid, had been omitted. The formation stopped taking the fluid and the charts showed pressure jump from 6,800 pounds to 8,600 pounds. Then all pumping should have ceased. This sudden pressure, forced against the bottom of the packer caused the packer to pump up the hole, and frac fluid and sand were forced on top of the packer. The casing chart of this well showed the 1,500 pounds pressure had jumped to 2,800 pounds pressure. They still pumped for awhile.

We were unable to pump down the tubing and out the casing, or down the casing and out the tubing. Everything was blocked with sand. Then, we got wash pipe to clean out over the packer. Finally we got the tubing, with a ruined packer, out of the hole. The packer and 6,000 feet of tubing had to be replaced. New tubing was purchased and installed to produce the well. It would not flow. A pumping unit and rods were purchased; the well pumped a small amount of oil.

This well has produced approximately only 600 barrels of oil in a two-year period. This mistake cost in excess of $180,000, and it did not produce in an economical manner. The investors were charged for this mistake, and a well was ruined.

How much potential profit did the investor lose? Was it thousands or hundreds of thousands? Who was responsible for the mistake? Were the investors ever informed about what honestly happened?

Incompetent Consultants

Once, I was questioned by a superior about a similar case. An engineer, fracing a well, saw the pressure gauge quickly jump from 6,200 pounds to 7,200 pounds. Pumping stopped

Incompetence

while they sat discussing alternatives. Finally the engineer decided to again try pumping into the well at a rapid rate. Surely things had settled out in the tubing. He pumped at too fast a rate and ruptured the tubing, which in turn forced sand on top of the packer. This little error cost investors about $76,000. The well was never as good as it might have been. A third of the frac job was dumped in the mud pits. This particular well produced about 110 barrels of oil a day, but with the correct frac procedure could have produced at least 170 barrels of oil daily. This represents a 60-barrel a day difference in production. For a 100-day period, 60 barrels a day production amounts to 6,000 barrels of oil in lost production. (6,000 barrels × $30 per barrel = $180,000 lost.) To my knowledge, none of the investors in this well made inquiries about this.

I've stated that the size of the pipe and the pressure it is designed to withstand are factors that determine the frac procedure. Also, if enough volume rate per minute is not obtained pumping down the tubing, the tubing is pulled. Then, the frac job is pumped down the larger pipe, the casing.

On one occasion I'm acquainted with, a consultant recommended a particular frac procedure, running down the casing. The service company started pumping the frac fluid and the sand mixture. About halfway through the frac job, the pressure, stabilized at 4,000 pounds, suddenly dropped to 1,800 pounds. They continued pumping and finished the frac job with this low pumping pressure.

The rig crew knocked the service company lines loose and began placing the tubing and packer in the hole. The well was drilled to a depth of 7,800 feet. When the completion rig tool-pusher called us, it was late in the afternoon. There were problems. They had lowered the packer down to 3,200 feet and it was stuck; it would not move up or down. The consultant had left the site.

The casing had ruptured, and as pumping continued, the fluid and the sand went onto the outside of the casing, forcing it to collapse. This pumping operation should have been stopped when pressure dropped from 4,000 pounds to

PROTECT YOUR INVESTMENT

1,800 pounds. That should have been an indication that something unusual had happened.

We finally got the packer loose and back out of the hole. Then we went in with a mill, milled the casing, patched the casing — the cost $90,000. The well would not flow. However, the well was shut down in order to gain an overnight pressure buildup. It rose to 2,200 pounds. A pumping unit and rods were purchased and installed on the well. While the well sometimes pumped oil, it was never satisfactory after that incident. This well was ruined by incompetence. Half of the frac job was lost, at $30,000. One questions the qualifications of the consultant.

I recall one more event involving well equipment in a frac procedure. Again, a consultant was involved. Employed by an oil company, I was in the field, checking wells, location, and drilling rigs. One morning I called the office and was informed the drilling manager wished to speak to me. He told me to go to a particular well location as soon as possible. They were trying to frac a well and could not get their pump lines tied into the wellhead. After about a 60-mile drive over fairly bad roads, I arrived to find the crew halfway through the frac job; the consultant was not on the location. I asked the service company frac master about the problem.

Apparently, the service company had been given the wrong size for the connections on the wellhead. Finally they had managed to hook onto the wellhead, but it had only a 3,000-pound working pressure. *They were pumping with 6,600 pounds pressure* on the wellhead. It could have blown up at any time.

The frac master said his people had found a 3,000-pound wellhead and so informed the consultant. The consultant gave a go-ahead to proceed fracing the well. The frac master wrote a report. Needless to say, the consultant's services were no longer required.

This mistake originated in the engineer's office, then was carried out in the field by a consultant. What if the wellhead had blown and began flowing wild, out of control — or blown up, ruining the well and possibly maiming or killing

—106—

Incompetence

Preparing to complete.

PROTECT YOUR INVESTMENT

someone? It's my belief that investor(s) should be due a sizeable sum of money in cases such as this where incompetence has created problems and losses.

Well stimulation can mean the very life of the well. It is most important to keep an eye on the results of the stimulation procedure. Just because a frac job goes sour and is costly does not necessarily mean the investor should lose.

It is a relatively easy process to check into the records that are maintained by the companies involved.

Investigate for Competence

To begin an investigative study, use the records of the frac job service company. (Such companies as Dowell, Haliburton, Western, and B.J. Hughes, etc. are competent and reputable.) These companies are in the well stimulation business and other defined aspects of the oil and gas service work. They maintain written records of each job performance from the proposal stage to job completion. They follow up on final procedures to observe and record the results. Their records include everything between the time of arrival on location until the departure time from location; they record any deviation or unusual happening during the frac procedure. Service company personnel aid in every possible way within their capabilities. I've had to lean on them from time to time, but I've had to gain their knowledge in my own field of work. Years ago, I knew nothing of frac procedures.

I firmly believe that if oil company engineers will give all the pertinent facts to the service companies on any well stimulation job to be performed, they can together work out proper procedure. Service companies have the necessary employees, the combined knowledge, experience, and equipment to evaluate a job accurately and to complete it efficiently. If a qualified investor calls a service company office, he can be furnished with the desired, necessary information. Their services are expensive, but they are

Incompetence

backed by research and development programs, and their equipment is operated by qualified personnel. Indeed, their records list the employees brought onto each job program, so it may be possible for the investor to consult directly with personnel on the scene.

The workover or completion rig on a well stimulation project also keeps records which are readily available to the investor. To obtain access to these records, however, in all probability you will have to first call the promoter through whom you invested. It is advisable to ask for the name and telephone number of the workover or completion rig company. The rig companies keep their own records, and they too are accessible to a potential investor. These records list the personnel of a particular location job. In addition to the personnel, their records indicate the performance duties of each on a daily basis. Rig company records compared with service company records can establish a probable cause of a costly problem, if the investor suspects there may be incompetence.

Most people involved in a mistake want to be cleared of wrong-doing in order to maintain their good names and to continue to work in good standing.

Get the facts first-hand, not through a second or third party. Always remember: the people in the field know more about what is taking place on a well than a person behind a desk does.

Chapter 8
DEFINING RESPONSIBILITIES

The investor should be concerned about all facets of a drilling venture. In my opinion, his concerns should include determining at the outset who is to be responsible in the event of errors, overcharges, damages, mistakes of incompetence, and major disasters. A part of his package should be a clear outlining of responsibilities and powers of decision-making.

In some cases the investor might wish to be consulted on decisions that might involve the total loss of his investment. In other cases, where rapid or immediate judgments have to be made, his involvement might work to his disadvantage.

Whatever you, as an investor, choose to do, you should certainly decide at the outset.

Irresponsible Cuts in Equipment

One potential investor, having examined all aspects of a drilling venture, plans to invest $50,000. On the A.F.E. form is an item listed as "surface pipe." This pipe is set in the ground to protect the fresh water supply; and, at the same time, it furnishes a good anchor facility on which to attach the blowout preventer that is used during the drilling.

In the State of Texas, the water control board sets up rules and guidelines as to the amount of surface pipe to set in the ground in a given area. These guidelines vary from area to area. The amount of pipe set depends on the depth of the

PROTECT YOUR INVESTMENT

well to be drilled and also at what depth the lowest freshwater sand will be encountered.

This specific A.F.E. states 1,800 feet of 12 3/4-inch surface pipe is to be set. The oil company checks with the water control board and files the correct footage permit for the specified area. However, the oil company decides to make extra money and sets only 500 feet of surface pipe, although they send a signed form to the water control board stating 1,800 feet of 12 3/4-inch pipe has been set and cemented. (Assuming 12 3/4-inch pipe sold for $9.50 per foot, $12,350 is saved on the 1,300 feet of pipe not set.) They also save on cement, and time is gained by not drilling the larger sized hole. (It takes longer to run 1,800 feet of pipe in the hole than 500 feet.)

Overall, they saved about $16,000. The investors, you realize, never see the pipe shortage or the profit. What is done is not part of the investors' agreement.

In the process of drilling this well, an oil and gas zone is encountered around 7,000 feet with abnormal pressure (due to a water drive in the formation). The well starts kicking back or blowing back. The rig crew must act quickly. They pull up and close the blowout preventer, and the well does not go out of control. The crew then starts building up the drilling mud weight in an effort to keep the well under control and to resume drilling operations. This procedure goes on for three days. On the fourth day, a farmer living nearby has gas and oil in his drinking water. Soon after this, another farmer reports finding oil and gas in his water. Suddenly, there is oil and gas in the drinking water of five different farmers' homes. The water supply for the livestock is also affected. The farmers will be certain to report it, and the source of the problem will be located.

The Water Control Board will see a logging procedure that indicates at what depth the bottom of the surface pipe is located. The damage suits will proceed and an investor wonders if he or the drillers will be allowed to attempt to correct the problem in the well bore, or to plug the well as is and leave the scene. Damages can run into the millions of dollars!

Defining Responsibilities

Will the insurance cover it? This is doubtful, because the proper amount of surface pipe was not set; also, a false affidavit was signed and sent to the Water Control Board. This sort of situation could break a small oil company.

Investors should find where their responsibility begins and ends. It is advisable to have a listing of responsibilities in the agreement between the investor and the oil company or the promoter, signed by all before drilling operations begin. If something such as the above occurs, does the investor lose all his money? Does he get sued? Where does he stand? The investor may be in the position to sue the oil company or promoter, if there is any money left.

Blowouts

I suspect there must be more drilling rigs buried in Jackson County, Texas, than any other similar area in the United States. Offhand, I remember twelve buried rigs. There were probably more, but I was not aware of all of them. In that area, there were a good many shallow gas producing zones that had a water drive on them, and the zones or strata consisted of pure sand (ready to flow). Drilling rigs were moved in, and contractors who did not know the area did not carry enough drilling mud weight, so blowouts occurred. When a blowout occurs and the blowing well brings sand and water, it is usually a complete disaster.

To help visualize what takes place, the sand and water is forced out of the ground with a great pressure. This is considered a sandblasting operation; with perhaps 1,800 to 2,000 pounds of pressure, it cuts everything in its path. If a situation of this kind exists, progress is not quick enough; it is hopeless.

Here's a specific case: I met a tool-pusher on a drilling rig late one afternoon. He asked for a bulldozer. I asked how deep the company planned to drill. He told me 7,800 feet. It would take him just a few days. He was drilling extremely fast, and he did not plan to weight his drilling mud until he

PROTECT YOUR INVESTMENT

reached 5,000 feet. I wanted to know if he knew where he was drilling.

Of course he knew! The reason he was drilling so fast, he explained, was because they were cutting or just washing a lot of sand and using light-weight mud. Heavy weight drilling mud slows the cutting action of a bit, similar to that of a heavy lubricant.

The next morning I returned to the wells I was in charge of. Immediately, I saw something blowing in the air, coming from the area where the tool-pusher's rig should be. The well apparently had started blowing out about midnight, when the crews were changing shifts. The top of the derrick was cut out; the drill pipe, standing in the derrick, had been cut in half. The substructure was cut to pieces, and the blowout preventers were wobbling around. A total disaster for an eight hour period, continuing for some time afterward. Absolutely nothing was saved; millions of dollars were paid for damages, and a gas well was lost to the air.

Another example of irresponsibility involved a bad welding job. One Christmas day a good many years ago, I was responsible for changing gas metering charts on some wells. When the day set for this procedure arrives, it must be done whether it is Christmas, a birthday, a wedding day, etc. While I was there; a chart changer for the pipeline company to whom we were selling gas arrived. He asked me if I could contact a man who worked for an oil company about five miles away. I had a couple of telephone numbers for the man, so I told him I would. It seems they had completed a gas well the day before, Christmas Eve. (In those years, drilling rigs were most often used to complete wells.) The wellhead was secured and ready to tie into the pipeline, but a small pinhole leak had developed in the surface pipe. However, the crew had finished, laid down the derrick, and then had taken off for the Christmas holiday.

After several attempts, I contacted a representative of the oil company and explained the situation. He told me that none of his crew were available and asked me if I would take a look at the leak, offering to compensate me for my time. I

Defining Responsibilities

Blowout preventers on a workover job. Equipment is designed to close top of hole, control release of fluids, permit pumping into hole, and allow movement of drill pipe.

told him it would be awhile before I could get back to him, because I had to drive about ten miles over rough roads. He said he would await my return call.

When I arrived at the well, the pinhole had grown to about one inch, and sand and water were coming out. I called the man and reported the conditons. He said he would call the drilling contractor and be on his way; would I mind getting a dozer and driver, a crane and floodlights to the well as soon as possible? Also he asked me to drag whatever I could out of the way. It was Christmas, but I was able to get assistance from a contractor from Victoria, Texas, about 40 miles away. At nightfall I had finished my

PROTECT YOUR INVESTMENT

own work, and I went out to look over the well. They had managed to save two pumps, a dog house*, and one generator. The sand and water had cut halfway around the pipe and had cut some of the legs on the substructure. By the next morning, the wellhead was cut off, and the rig itself was covered with sand. Salt water, sand, and gas were blowing over 300 feet in the air. The well was never brought under control. By the time it stopped blowing, 2,000 acres of riceland was covered with salt water. The rig was a total loss.

This had begun with a bad welding job. However, the gas should not have been channeled up into the surface pipe, and the drilling rig should not have been left there unattended, under the circumstance.

Look at this from an investor's point of view: A good gas well one day, and the next day nothing. Did the oil company have a qualified welder perform the job? If the company's insurance will not pay damages, who is next in line to be sued? Is it the investor? Where does the investor's responsibility stop? Almost all of the people with whom I conferred about this well blowout were of the same opinion: the drilling rig hampered the progress of getting the well back under control. Was the drilling contractor instructed by the oil company to leave the rig on location or to move it off? A blowout of this nature depletes billions of cubic feet of gas. The investors had the gas, then lost it. Are investors entitled to a huge profit for gas lost to the air? An investor must determine where he stands. A well-planned venture can create a millionaire, and a venture such as the one described can break an investor.

Used Equipment

I can relate another blowout operation that involved equipment. A promoter obtained a mineral lease that stood between two producing gas wells. I don't know how he obtained the lease; he had no money of his own. A well was drilled about a hundred feet from the shoreline on a South

Defining Responsibilities

Texas bayshore. The wellhead (christmas tree) was installed, and the well was brought in. The well was wide open, blowing good; and the crew was in the process of cleaning up, which was moving quickly. The trouble began when it was decided to shut the well in and install a choke in the wellhead to control the hard blow. None of the valves on the wellhead would close. The promoter had purchased a used wellhead; no one tested the valves for proper function. There was no way to tie into the wellhead so that mud could be pumped into the well. The rig had to be removed. The well cut the wellhead to pieces; water, sand, and oil spilled into the bay. Consequently, the State of Texas, the Parks and Wildlife Agencies, and countless others went looking for the promoter.

The point is that the promoter installed a used wellhead. He had a good well, but lost it, and the damage was extensive. Used equipment will save money, but only if it works properly. The investor should be informed if used equipment is to be installed. If used equipment is installed but investors are charged for new equipment, guess who has been cheated!

Quick Decisions

To bring any well under control after a blowout has occurred, time is of the essence. Knowledgeable, sincere people must initiate the work.

The sooner a blowout is brought under control, the easier it is for the wild-well fighters to save the equipment on the well and the hole pipe. If a well blows or burns over a lengthy period, the condition of the equipment and pipe deteriorates at a rapid pace. When extreme damage occurs, there is no way wellhead control equipment can be attached. Control equipment placed on the blowing well and finally shut will cause the pipe to rupture, because of heat crystallization on the walls of the pipe, worn thin by washing sand and water under pressure. Equipment will

PROTECT YOUR INVESTMENT

bear just so much abuse under extreme pressure.

Another thing to remember is that the longer the delay in getting a blowout under control, the less will be the profits for the investor. While a well blows out of control, whether oil or gas, the investors' profits are going into the air and onto the ground.

Obviously, a blowout is a most serious matter. Suppose $50,000 is invested in a well on a 200-acre lease tract. There are wells on three sides of the lease that have been producing for a short period of time. Our one well blows out at the same depth of those nearby wells that are leased to others. It is out of control for three weeks, and, in time, the producing wells on these adjoining leases stop production. Has our blowout depleted their reserves? Will we — or our oil company — be sued? Of course, our well should have never blown out. But wells blow out frequently; it's an operational factor that investors should be aware of. We should be informed before making our investment what our position will be if something like this should occur.

The longer a well blows out of control, the higher the cost of damages. This cost factor involves the immediate area of the well site and also the surrounding land, rivers and streams, public roads, utilities, and homes. In one South Texas locale a railroad had to be moved due to a well blowout. Today the long curve in the rail still stands.

At what time should the wild-well fighter be called to go to work on the blowout? Does a drilling rig crew have authority to call the fighter? Does the operator or the oil company in charge of a drilling operation? Do we have to go through the insurance carrier to call the fighter?

Scores of inquiries have not given me the answers. When it's decision time, a fast and accurate one must be made. One insurance adjuster told me if the wild-well fighter is needed, get him. But who decides if one is needed?

Another adjuster told me the only cause of any blowout is the mud weight. This may be an educated guess. The mud weight controls the pressure that will be encountered while drilling, *but* the density and the viscosity of the drilling mud are controlling factors also. Drilling mud can be

Defining Responsibilities

weighted with sand and be worthless. The density of drilling mud, in a brief description, is the compactness of the mud itself. The viscosity of the drilling mud is the property of the fluid that determines the rate of flow and is closely related to its internal friction.

The driller on one rig had his mud in top shape. He began to pull the pipe out of the hole — a task that required all the crew. No one noticed while pulling up the drill pipe the drilling mud oozing over the top and flowing back into the mud pit. The crew pulled about ten stands of pipe, and the well started kicking or blowing back. Before anyone could close the rams on the blowout preventer and stab back into the drill pipe, the well blew out of control. How could this be, if the driller had perfect drilling mud?

A bit has, most often, three jets or nozzles in the bottom. In the preceeding example, the driller was not moving the volume of drilling he thought should be pumping. He suspected that some of the jets or nozzles might have become plugged; that the bit was dull, had become balled, and the bearings ruined. To determine if this had occurred, the crew pulled the pipe from the ground. The bit acted like a swab and pulled mud to the surface, leaving very little volume of mud below the drill bit. Obviously, the drilled hole was not kept filled with drilling mud to its top. When the mud weight or volume dropped to a lesser weight than the pounds per square inch encountered in the ground, the well blew out.

From my experience and knowledge of drilling mud and its properties, I am of the opinion it is not only mud-weight itself that is a contributing factor causing blowouts.

There is a great deal of misunderstanding about the mud weight in respect to the size of the hole that is drilled. The mud weight versus the diameter of the drilled hole makes no difference. The drilling mud weight is designed with density and viscosity to control the pounds per square inch of pressure exerted against any size surface area at any depth to be drilled. A two-inch hole drilled to 10,000 feet would still require the same mud weight as a twenty-inch hole drilled to that depth. It is the depth of the hole plus the

PROTECT YOUR INVESTMENT

volume of proper weight drilling mud required to fill the drilled hole to the surface that are the controlling factors in preventing a blowout.

The Insurance Representative

How can investigating a well blowout make you money?

Consider that you have invested in a well, and suppose it blows out of control for three days. All of the persons involved with the drilling are at the site, including the drilling contractor, the oil company president, and the insurance representative.

The insurance man has already suggested that it's time to get a wild-well fighter on the scene. However, he wants to use a specific fighter. No one else will do, he emphasizes.

This particular fighter is busy and cannot arrive at the site for another day. Four days (ninety-six hours) have passed by the time the fighter arrives. Immediately he surveys the site. The equipment and the well pipe on top of the ground have been cut irreparably by the sand, water, and pressure. There is no need to attempt to tie any kind of wellhead control equipment onto the battered ones. Another day (the fifth) of study brings a solution.

He gets the rig set up and drills a directional hole into the blowout well in an effort to bring the well under control. Operating this plan results — two weeks later — in stopping the blowout, but the wellhead equipment was destroyed. Remember, it was the insurance company person who wanted to use only one particular fighter. They had to wait a full day for him and this might not have been soon enough to save the well.

That delay in well control procedures was caused by one person, the insurance representative. Was it the delay that was fatal to the wellhead equipment? Could the well have been brought under control if the wild well procedures had begun one or two days earlier? Where does the responsibility rest? In this example, the insurance company representative assumed responsibility. For that

Defining Responsibilities

reason, the company should be required to pay for damages caused by delay.

In my opinion, investors should determine at the outset what procedures are to be followed in the event of such an emergency as this. Make sure responsibilities are assigned and that the responsible person understands the importance and the consequences of his decisions. It is the investors' money that is at risk, and it must be handled responsibly.

Chapter 9
INVESTING POINTS AND QUESTIONS

How long should the investor allow the promoter access to his money? Each investor has specific ways to handle money, for obvious reasons. What are some guidelines that will prove to be beneficial?

Tax Incentives

First, we recommend seeking the advice of a tax expert. Internal Revenue Code Section 172. Net Operating Loss Deductions, known as NOL, is now in effect. The law provides for the carry back of net operating losses for three years prior.

The tax provisions pertaining to the oil and gas business are extremely complicated. For information on the tax consequences of oil and gas investments, investors should consult their tax advisors with reference to their own tax situations.

The Internal Revenue Code permits 100 percent of the intangible drilling costs of drilling an oil or gas well to be tax deductible in the year those costs are incurred. Intangible drilling costs include all costs that have no salvage value and generally amount to 60 percent to 80 percent of the total cost of drilling a well.

Drilling costs associated with producing properties also enjoy a tax benefit. The percentage depletion provision currently permits 18 percent of gross income, not to exceed 50 percent of net income, to be received tax free. This rate will be reduced to 15 percent by 1985.

PROTECT YOUR INVESTMENT

Oil and gas exploration is an investment opportunity capable of countering taxation and inflation. The investment is largely tax deductible, and a portion of the income from producing wells is received tax-free. I am of the opinion that, by investing in oil and gas exploration and development, an individual has the opportunity to turn taxable income into an investment that creates capital assets producing income, partially tax-free, for many years.

When Do You Turn Over The Money?

We know each drilled well is as different a happening as the different ways an individual invests his money. Each well drilled and each investment have similarities to others, but no two cases are the same.

My caution is not to let the promoter have access to your money until such time as the money is needed to pay the outstanding bills charged to the well in which you are investing.

Suppose an investor purchased a 4 percent interest in a drilling venture on March 15, 1982. A 1 percent interest in the venture is selling for $10,000, a total of $40,000 for the investor's 4 percent working interest. The promoter does not start drilling the well until January 15, 1983, ten months in which the investor could have his money working for his own purposes.

Let's assume it will take 20 days to drill the well and reach the logging point. The well logs look good, and casing is to be set in the well. This adds another 5 days because the drilling rig must be moved from the location, after the casing has been cemented in the well. Then a completion rig is brought in to complete the well. This adds another seven to ten days to the time element. Now, we have eleven months that the promoter held the investor's money. About this time, the promoter begins to receive the initial invoices.

He has held the investor's money for one year, during which time he sold 80 percent of the 100 percent working interest for $10,000 per 1 percent interest. This gives the

Investing Points and Questions

promoter $800,000 of investors' money. Then, suppose he keeps the remaining 20 percent of working interest for himself. We could take for granted that he would put $200,000 of his own money into the well.

Not so! Assuming the investors placed their money in March, 1982, the promoter most probably deposited the $800,000 in a bank account where it drew 14 percent interest for that year. Fourteen percent interest on $800,000 produces at least $112,000 in a year. Therefore, the investor has paid for 11.2 percent working interest in the well that will be free to the promoter, who will have to pay for only 8 8/10ths percent working interest, which amounts to $88,000. Actually, the promoter paid $4,400 for each 1 percent working interest while the investor paid $10,000 for each 1 percent.

This whole operation can very easily snowball. The chairman of the board with a small independent oil company is talented in money maneuvering, but he does not know much about the drilling business. Consequently, he hires people who do know drilling operations. For example, this executive places people on the board of directors who have interests in the pipe business, the wellhead business, mud logging, and the banking business. It's understood that each well this company drills must use the companies that the board members are associated with, so their extra money will be earned — even if the higher prices are charged to the investor. This same company also has two marketing agents, plus a journalist for promotional purposes. It is their intention to drill wells that cost around a million dollars or more. No shallow, cheaper wells.

The company has its mineral leases, and work begins. In the first six months, eight wells are promoted to drill. The company is established with $7,000,000 by the chairman and board of directors. Assume the company promoted eight wells in six months at $1,000,000 each, and keeps 20 percent of each well, it is entirely possible that they have banked $6,400,000 of investors' money at the end of that six-month period. In the seventh month, drilling begins on the first well.

PROTECT YOUR INVESTMENT

It's a safe guess that the company's expenses would run around $200,000 a month for the first six months. This being a new start up, this amounts to $1,200,000 total for that interval. The company has $6,400,000 for investors, plus $7,000,000 of it's own money, minus $1,200,000 expenses, which leaves $13,400,000 in one or more banks.

Let's take a look at the next year's progress. If two wells are promoted and drilled per month, that total is 24 wells a year. Say the company averaged $100,000 from each well in interest on the investment of the investors' money for one year. The company could make $2,400,000 interest and, at the same time, be getting revenue from the oil and gas production.

Do you see how someone else can make use of your money? Keep your money working for yourself until it's needed to pay debts incurred on the well you have invested in.

Alas, there are so many more ways for investors' money to be abused. Here is another example. A very small oil company is formed and begins picking up small investors. This company will drill two or three wells and pay their bills. Then it will drill a well and not pay part of the invoices. During this brief time of operation, the oil company has formed another company to operate part of the properties. (Let's not forget the original company is still searching about for investors.)

The original company will also pay operating capital to the second company and will have itemized bills to prove it. Suddenly, any oil and gas properties the original company owned are sold. It stops paying its bills. One can sue them, but they won't have any money. In addition, one will not find the owners. But we can be assured of one thing: they will start up a business again later — under another name.

An important topic for discussion is the time element of investing for tax purposes. In the earlier example we placed investors' money in oil company's hands in March, 1982. However, the money was not used until 1983. The investor may have received a better tax break if his money had been

Investing Points and Questions

spent the same year it was invested. An aware investor must be familiar with the tax loopholes and benefits.

How Far Will Your Money Go?

With respect to proper investment points of a drilling operation, there are different stages in a venture in which investment may be made. The promoter may ask the potential investor if he wants an interest through the logging point, through setting casing point, through the completion point, or through the sales point.

1) *Through the logging point:* The investor may invest a specific amount of money into the drilling venture to pay the drilling and logging of the well. This is the highest risk investment in a well.

2) *Through setting casing point:* The investor may invest a specific amount of money into the drilling venture to pay the drilling, logging, and setting of the production casing.

3) *Through the completion point:* The investor will invest a certain amount of money into the drilling venture to pay the drilling, logging, setting the production casing, and the completion of the well.

4) *Through the sales point:* The investor will invest his money into the drilling venture to pay the drilling, logging, setting production casing, completion of the well, and oil production facilities if needed, or a metering and sales line for gas, if such is needed.

Are You Paying Someone Else's Way?

Investment procedures seem to fit in with the investment points. I could not recommend investing in a well at all if you are approached by a promoter who has a good mineral lease and who wants to drill a well to a total depth of, say, 5,000 feet: he offers to sell the investor the drilling from 2,000 to 3,000 feet.

PROTECT YOUR INVESTMENT

That's all the investor pays for! A thousand feet of drilling.

The bottom line is that the investor will receive only the production that comes from 2,000 to 3,000 feet. The promoter goes to another investor with the same song and dance except this time it's 3,000 to 4,000 feet. And when he gets his last investor for the last 1,000 feet, he is ready to drill.

When Is A Dry Hole Not A Dry Hole?

I actually saw this happen in Jackson County, Texas, some years ago. A promoter put a drilling venture together in this fashion. However, he promoted a total depth of 4,000 feet. There are a good many shallow gas zones in the Jackson County area. They ran the open hole logs on the well, and the promoter then plugged the well. In the well plugging procedure, they set a cement plug up in the surface pipe. The drilling rig was then moved out.

Having plugged the well, the promoter called the investors and explained that they had drilled a dry hole. However, there were some small gas carrying zones, but they were wet. (The wetness means the zones were carrying too much water to be productive.) He informed the investors there might be a company that would buy the lease at a $10 per acre profit. This mineral lease contained 420 acres. This was a promoter's way of "giving his all" to his investors and, of course, they would get a return of some of their money. But this promoter was on his toes; he had now become one of the "good ole boys," he had covered up his tracks for the next go-around at this lease.

The investors never looked at the well logs, certainly a grave and costly mistake. They just took the promoter at his word. He, on the other hand, formed a new company by filing an assumed-name affidavit and then sent a check to the investors in the amount of $4,200, representing the $10 per acre *profit* on the 420 acre lease.

Investing Points and Questions

This was not a profit at all! The promoter bought out the investors because he wanted them out of the picture completely. By this procedure, should questions arise — and they will — he simply states that he bought them out for $10 an acre. The investors believed this promoter was the best buddy in the world because he didn't keep any of the profit from the sale of this mineral lease.

The promoter had about a year remaining on the mineral lease before the agreement expired. For a ten-month period after the original well had been plugged and abandoned there was no drilling rig on the place.

Before the mineral lease expired, the promoter moved a drilling rig right back on the original well and prepared for a re-entry. He drilled out the two cement plugs and attempted to make a well. Why? There were six or seven potential gas zones in the original hole drilled. The well logs had been released and made public. Hunt Oil Company, Mobil Oil Company and Siboney Petroleum Company had picked up leases surrounding this mineral lease. Along with South Texas Oil and Gas Company taking up a nearby lease, these companies had seen the original well log and liked the reports.

The drilling rig began operations. However, the promoter did not tell the drilling contractor that several gas zones had been encountered below 2,400 feet. The rig drilled out the first plug in the surface pipe. Nothing happened. The drillers washed on down to the 2,000-foot cement plug, and all of the drill pipe was blown out of the hole. The derrick was torn to pieces by drill pipe being thrown all over the property. Luckily, no one was killed, but this blowout bankrupted the drilling contractor.

From 2,000 to 4,000 feet of this re-entry hole several of the gas zones had become as kegs of nitro. The drilling mud that had been used and left in the hole by the original drilling riggers settled out, which means that the heavy drilling mud additives had settled to the bottom of the hole and water had risen to the top. In itself, water will not hold gas with as much pressure as was involved in this well. In

PROTECT YOUR INVESTMENT

Ninety percent of American oil is produced by artificial lift.

Investing Points and Questions

the drilling of a well, the use of clear water is begging for trouble. Needless to say, everything went sour for the promoter.

I was told of this drilling venture by one of the original investors. When a well blows out, it is oilfield news and will most assuredly make headlines. The investor had read about the blowout in the newspapers and came to the fields to look the situation over. He told me he had learned a lesson, and he hoped the promoter had been taught something also. I doubted seriously if the promoter wanted to be taught any lessons. He would more likely think up something else to try.

By the way, there were about thirty good wells made around the region of this one blowout, which produced for years. No, the promoter did not have any producing wells in this field. It's safe to say he discovered only one well — with a big bang.

The reason I mention investment points of a drilling venture is because all ventures do not have to be financed in the same fashion. I have known people with good leases and the money to drill the wells. They may have needed money to complete their drilled wells. I have also seen gas wells shut because no money was available to lay lines and get the gas into a sales line.

One man I know had a good-sized lease. He drilled and completed a good gas well. With all of his money gone, he had to sell part of his lease to pay for the pipeline. So, there are as many different deals as there are investors who make them.

Are There Any Loopholes?

Working over a well is another operational expense about which an investor should learn. We know a gas well or an oil well will not last forever. At some point any oil or gas well will become an operation no longer economical. When this point is achieved, it may be possible to rework the well and resume profitable production. However, we must keep in

PROTECT YOUR INVESTMENT

mind that this can become an expensive undertaking.

The informed investor, aware of workover procedures can require the cost in writing for his written approval. There is still another method of handling the process, one that is too often abused.

Let's suppose there are ten investors in one well. In the agreement with the people who will operate the well, there is a paragraph stating that if workover costs exceed $10,000, the operators must have written approval from all investors. But there is a loophole in the clause. The well operators worked the well over four times within six months. Each workover cost only $9,000, but in fact they spent a total of $36,000. The well was not in any better shape than before the workovers began. Therefore, investors, check the procedure for handling workovers, or profits could be eaten up.

Audit to obtain accountings of how and where the money is used. An investor, placing his money in a drilling venture, receives a signed copy of the A.F.E. from the promoter or the oil company he represents. Investors can ask for an itemized statement of expenses and compare the statement against the A.F.E. If irregularities are discovered, an explanation of the cause is necessary. Investors are not expected to run a drilling operation, but they are entitled to know their money is handled properly.

Further Questions To Ask

Investors can require answers to each of these questions:
— Is the mineral lease good and free and clear of any liens?
— Are 100 percent of the lessors signed into the lease agreement?
— Do the lessors hold only the mineral rights, with the surface land owned by another party?
— What is the original cost of the mineral lease and who are the lessors?
— What percentage of the royalties did the lessor keep?

Investing Points and Questions

— What percentage of the mineral lease has gone to overriding royalty interest, and who holds the interest?
— What percentage of the revenue is paid to the 100 percent working interest?
— Who will the other working interest investors be? (Demand a list.)
— Is the investor purchasing an interest in one well or the entire mineral lease?
— Where did the production figures on adjacent wells come from, and are the figures documented?
— Are the people promoting this mineral lease familiar with this particular area?
— Is the area the mineral lease is located in cut with a great many faults?
— How were the reserves calculated?
— How much of the reserves may be recovered in an economical manner?
— How many years, and at what declining rate, will it take to recover the anticipated reserves?
— What will the proposed price scale of the oil or gas be during the recovery period of the oil or gas?
— Will the investors be furnished copies of the run tickets for the sale of oil from the lease? (Obtain copies for personal records.)
— If a gas sales pipeline must be laid, what percentage of the pipeline will the investor own; how will future gas put through the pipeline be paid back to the pipeline; and will all gas be metered at a final sales point? (Obtain documented answers for personal records.)
— Will the investor be charged for incompetent mistakes that produce a cost overrun? (Obtain documented statements.)
— Where do the investors' cash responsibilities start and end?
— When buying producing properties, where did the property originate?
— Where are the documented production reports for the producing property that is offered for sale?

PROTECT YOUR INVESTMENT

— What is the sales agreement between broker and seller:
— Are there any liens on the producing property that is being offered for sale?
— Are there any pending lawsuits against the producing property that is being offered for sale?
— When should the investor let the promoter of the oil company have access to the investors' money?
— What are the tax advantages to the investment of money?

Finding the Good Guys

As in any field of endeavor, the bad guys of the oil business will be noticed and give the business a bad name, while the good guys play it straight and often go unnoticed. The good guys in oil will eat, sleep, and talk the oil business twenty-four hours a day, seven days a week. It is in their blood; they like it and they know how to take the hard knocks. They are very competitive, but they will also help each other.

Any time a promoter approaches an investor and states, "I promise you this well will make 200 barrels of oil a day," or "This well will produce 1/2 million cubic feet of gas a day," that is the time to beware. There is no way to tell how much oil a well will make until you have oil going into the tank or gas going into a sales line.

The good guys will tell you that they *think* this well *might* produce a certain amount, but then they will explain how they arrived at the answer, giving the facts collected for the venture. The good guys will explain any problems they think may be encountered in the drilling and completion of the well to the best of their knowledge. They want the investor to be aware. Why? Because the investor becomes a partner in the effort to make money and to keep on making money. The good guys are not interested in a one-shot deal and then goodbye.

Ask the person promoting the drilling venture how long

Investing Points and Questions

he has been in the business. Ask for references — personal and banking. Ask for the names of drilling contractors he has used and pipe suppliers. Then get on the phone and check him out. If these references give a favorable response, go one step further. Call your own bank and ask it to check out the promoter as well. Call other oil and gas companies and oilfield supply houses. If your promoter is a good guy, they will tell you. Don't be afraid to do this. If your promoter is reputable, he wants his good reputation known throughout the business. He will be glad for you to confirm it.

Another good source of information for the potential investor is the larger oil and gas companies. In a great many cases, the small operators work with the large companies, either in selling oil or gas to them or in financial ways. I offer an example of the way a small operator may associate himself financially with a large company. He may pick up a 2,500 acre mineral lease that looks promising, but he and his investors have only enough money for one well. Not wanting to lose a part of the lease, the small operator will go to a larger oil or gas company, show them his prospectus, and make a deal that will be mutually beneficial.

Should you wish to invest in the oil business, you may be surprised to find how really helpful total strangers may be. Despite the many pitfalls I have outlined for you, the majority of people in the business are good guys. If you write or phone these people with a genuine concern, they will be glad to help you obtain the information you need. The oil and gas business is a good one, with plenty of good guys who want their livelihood to have and keep a good name.

GLOSSARY

Abnormal Bottom Hole Pressure (BHP) — A reservoir is said to have an abnormal BHP when its pressure deviates appreciably from that of a saltwater column whose height is equivalent to the depth of the reservoir.

Accelerator — A chemical additive that reduces the setting time of cement.

Acid Gas — Undesirable and corrosive acidic components of natural gas, i.e., hydrogen sulfide and carbon dioxide.

Acid Fracturing — Process of opening cracks in hard carbonate productive formations by using a combination of oil and acid under high pressure.

Acidizing — A technique used in wells to increase permeability immediately around the well bore by injecting acid into the formation. Also used to clean the walls of the hole by injecting acid into the formation. Also used to clean the walls of the hole and screens or liners by circulating or injecting acid.

A.F.E. — Authorization for Expenditure

A.F.R. — Authorization for Retirement

Allowable — The common meaning in the industry is that volume of oil or gas production permitted (ordinarily on a per well basis) per day by the state regulatory body.

Amine Unit — A continuous system in which acid gas is removed from a hydrocarbon gas stream by an aqueous solution of monoethanolamine in a counterflow contactor, and the spent amine solution is reactivated in a stripping still with reboiler.

Glossary

Anchor — A device for holding, fixing, or fastening any object which may tend to change its position. For example, dead line, wire line, derrick anchors. Also, a length of tubing extending below the working barrel in a pumping well — gas anchor and mud anchor; a length of tubular goods extending below the packer when drill stem testing. (See tubing anchor).

Annulus — The space between the casing and the wall of well bore, or between two strings of casing, or between tubing and casing.

API — American Petroleum Institute

Artificial Lift — Any means of lifting liquid from a well that has ceased to flow.

Back-off a Joint — To unscrew the drill pipe, casing, tubing, or rods at a point above which it is stuck in the hole or to unscrew a joint of pipe at the surface.

Back Pressure — Pressure retained in a pipe line, vessel, or reservoir, by restricting the outflow of a gas or liquid.

Backup — The act of holding one section of pipe while another is screwed out of it or into it. A backup wrench refers to any wrench being used to hold the pipe. Backup tong is applied to the drill pipe tongs suspended in the derrick and used to hold a section of drill pipe while another section is screwed into it by use of other tongs. The backup man is the crew member who operates the backup tongs.

Backwashing — The process of washing the formation by injecting fluids into the formation, opening the well, and permitting the well to backflow.

Bailer — A long tubular vessel fitted with a bail at the upper end and with a valve at its lower extremity, used to remove water, sand, mud and oil from a well. When fitted with a plunger, to which the line is attached, it sucks the materials in as it is lifted, and is then called a sand pump.

Ball and Seat — The renewal parts of oil or water well pump valves. A ball and seat is a part of each standing and each traveling valve.

Glossary

Barite — Barium sulphate, a mineral used to increase the weight of drilling mud.

Barrel — One barrel of oil equals 42 gallons (U.S.A.).

Basket Sub — A fishing accessory run above a bit to permit recovery of small bits of metal or junk in a well.

Beam Hanger — A steel hanger attached to the end of the walking beam, used to suspend the sucker rods in a well.

Bell Nipple — A casing nipple installed in the top of the blowout preventer. The top end of the nipple is expanded — belled — to guide drill tools into the hole and usually has side connections for the fill line and the mud return line.

Bentonite — A colloidal clay composed of the mineral montmorillonite and having the property of swelling when wet. Owing to its gel-forming properties, bentonite is a major component of drilling muds.

BHP — Bottom Hole Pressure

Bit — A tool used in drilling wells which does the cutting at the bottom of the hole.

Blanket Gas — Gas, from an outside source, used to keep air out of a liquid storage tank.

Bleeder — A connection on a line or piece of equipment used for releasing pressure or draining off undesirable liquids.

Blind Flange — The flange used to close the end of a pipe. It produces a blind end which is also called a dead end.

Blind Rams — Integral part of the blowout preventer. Rams whose ends are not intended to fit drill pipe, but to seal against each other and shut off completely the space below.

Block — In mechanics, one or more pulleys or sheaves mounted to rotate on a common axis: any assembly of pulleys on a common framework. The crown block is an assembly of sheaves mounted on beams at the top of the derrick. The drilling cable is reeved over the sheaves of the crown block alternately with the sheaves of the traveling

Glossary

block, which is hoisted and lowered in the derrick by means of the drilling cable.

Blowdown — The material removed from a closed water system, e.g., cooling water system, boiler system, etc. to prevent the over-concentration of dissolved salts and precipitated sludge and to prevent foaming and carryover in boilers.

Blowout — A sudden, violent escape of gas and oil and sometimes water from a drilling well when high pressure is encountered and the well flows out of control.

Blowout Preventer (BOP) — A heavy casing head control filled with special gates of rams which may be closed around the drill pipe or tubing or which completely close the top of the casing if the drill pipe or tubing is withdrawn.

Boomer — A device used to tighten chains on a load of pipe or other material on a truck to make it secure.

Bore-Hole — A hole in the earth made by drilling.

Bottomhole Pressure (BHP) — The pressure at or near the bottom of a well. Bottomhole pressure is usually determined at the face of the producing horizon by means of a pressure recording instrument which can be lowered into the well. It may be calculated by adding the pressure at the surface to the bottom of the hole.

Bottomhole Pressure Bomb — A device consisting of a pressure element and a recording device usually enclosed by protective metals which may be run into a hole on a wire line and used to record the pressure at any depth in the hole.

Bottoms — The bottom liquid stream leaving a separation process in which a lighter or more volatile stream goes overhead.

Bottom Settlings (B.S.) — Sediment, earthy matter or inert organic matter which accumulates when crude petroleum is stored in tanks.

Bradenhead — A casinghead.

Break Out — Refers to the act of unscrewing one section of

Glossary

pipe from another section, especially in the case of drill pipe or tubing, while it is being withdrawn from the well bore.

Breaking Down — Usually means unscrewing the drill stem into single joints and placing them on the pipe rack. It is necessary to break the pipe down in order that it will be in lengths short enough to be handled and moved. Also called laying down.

Breather — A vertical piece of pipe, equipped with an open end return bend, located on top of an atmospheric pressure vessel to permit air to flow into or out of the vessel and thus prevent a vacuum or excessive pressure.

Breathing — Surging in the flow of gas from a well; also, applied to vertical motion of tubing during pumping operations. Escape of gas from stock tanks due to changes in temperature.

Bridge — An obstruction in the drill hole. A bridge is usually formed by caving of the wall of the well bore or by the intrusion of a large boulder or other debris.

Bridge Plug — A down-hole tool, composed primarily of slips, plug mandrel, and a rubber sealing element, which is run in and set in casing to isolate a lower zone while testing an upper section.

B.S. & W. — Basic sediment and water.

Bull Plug — A fitting made of a short pipe nipple, having one end welded closed or pressed in oval form and the other end threaded, used for closing any opening in a pipe connection or threaded outlet.

Burn Pit — An earthen pit for accumulating and burning unsalvable oil.

Bushing — A pipe fitting used for connecting a pipe with a pipe of larger size, being a hollow plug with external and internal threads to suit different diameters.

Butterfly Valve — A valve with a pivoted disc used to control flow.

Bypass — Usually refers to a pipe connection around a meter, valve or other control mechanism. A bypass line is

Glossary

installed in such cases to permit passage of fluid through the line while adjustments or repairs are made on the control which is bypassed.

Cage — The container for the ball in a ball valve such as found in the subsurface pump ordinarily used in oil production.

Caliche — A natural unconsolidated limestone occurring over large areas of west and south Texas, often near the surface.

Caliper Logging — An operation to determine the diameter of the well bore or the internal diameter of casing, drill pipe, or tubing. In the case of the well bore, caliper logging indicates enlargement of the bore. In the case of tubular goods, the caliper log reveals internal corrosion or other defects.

Cap — A fitting that goes over the end of a pipe to close it, producing a dead end. See "Bubble Tray."

Cap Rock — A comparatively impervious stratum immediately overlying a gas or oil-bearing rock.

Capping — The name given to a method by which the uncontrolled flow of a well is stopped or controlled.

Casing — A steel pipe placed in a well, usually cemented in place.

Casing Centralizer — A device that is secured around the casing to centralize the pipe in the hole and thus provide a uniform cement sheath around the pipe.

Casing Clamps — Two heavy steel bars (about 1 1/2" × 12" and several feet long in some cases) formed so that each is nearly a semicircle with ends projecting. The two bolted around the top of casing rest upon a foundation and form a support for the casing.

Casing Pressure — The pressure between the casing and tubing.

Casing Shoe — A short, heavy, hollow cylindrical steel section, beveled on the bottom edge, which is placed on the

Glossary

end of the casing string to serve as a reinforcing shoe and to aid in cutting off minor projections from the borehole wall as the casing is being lowered.

Casinghead — A fitting screwed to the top of a well casing. It is provided with side outlets for gas or oil, and provides a means for supporting the tubing.

Casinghead Gas — Gas found in an oil stratum and produced with the oil therefrom. Usually restricted to gas produced from wells classified as oil wells.

Cathead — One of the component drums or reels of the draw works assembly which furnishes power, through the catline and jerk lines, to lift heavy objects, make up drill pipe joints, break out drill pipe joints, make up casing joints and other similar operations.

Catwalk — The ramp to the side of the drilling rig where pipe is laid out to be lifted to the derrick floor by the catline.

Caustic Embrittlement — The intercrystalline cracking of metal due to accumulation of alkaline residue in a crevice.

Caustic Test — A test to determine the quantity of free sodium hydroxide in a caustic solution.

Caustic Unit — A system in which caustic is used to remove mercaptans and hydrogen sulfide from a liquid hydrocarbon stream, normally propane, LPG, or gasoline. The unit is "regenerative" if the caustic is continuously regenerated. It is "batch" if the caustic is periodically replaced without regeneration.

Cavity, Salt — An underground storage space, normally used for a gasoline, butane, or LPG, in a salt-bearing formation. It is made by drilling a well and dissolving the salt by circulating water.

Cellar — Excavation under the derrick to provide space for items of equipment at the top of the well bore. Also serves as a pit to collect drainage of water and other fluids under the floor for subsequent disposal.

Cementing — The operation by which cement slurry is

Glossary

forced down through the casing and out at the lower end in such a way that it fills the space between the casing and the sides of the well bore to a predetermined height above the bottom of the well. This is for the purpose of securing the casing in place and excluding water and other fluids from the well bore.

Centrifuge — An instrument for separating B.S. & W. from oil by means of centrifugal force.

Chain Tongs — A tool used in assembling or disassembling pipe or pipe fittings, having a lever with a serrated end, provided with a chain, to either turn pipe or hold it from turning.

Channeling — Ordinarily refers to bypassing of reservoir oil by gas or water. Also refers to bypassing of mud by cement during cementing operations. An undesirable flow condition existing when a fluid bypasses portions of a packed vessel or bubble tray due to improper packing or poor liquid distribution.

Cheater — Any additional length, usually a piece of pipe, added to a wrench for extra leverage.

Check Valve — A valve which permits a fluid to pass in one direction, but automatically closes when the fluid attempts to pass in the opposite direction.

Chloride Test — The Mohr test for determination of soluble chlorides in water by titration with a standard silver nitrate solution in the presence of potassium chromate indicator. Results of the chloride test are used to evaluate and control blowdown and to calculate makeup in boilers and cooling systems.

Choke — A valve or an orifice used to restrict flow and control the rate of production.

Christmas Tree — A term applied to the valves and fittings assembled above the tubing head at the top of a well to control the flow of the oil or gas or a flowing well. The christmas tree is attached to the top of the tubing head.

Circulate — The act of continuously pumping drilling mud

Glossary

or other liquid down the inside of drill pipe or casing and out through the annular space. This is done to condition the drilling mud and the well bore and to obtain cuttings from the bottom of the well bore. (See "Reverse Circulation")

Cleanout Plate — An opening in a tank through which the tank may be cleaned out.

Coat & Wrap — To cover pipe with bituminous, or similar material, and textiles for protection against corrosion.

Collar — A term used in place of threaded pipe coupling.

Column — A tall vertical processing vessel, e.g., fractionator, absorber, etc.

Completion — At a drilling well, that part of operation which takes place between the time the producing formation has been encountered and recognized and the well is ready to produce. A finished well, either an abandoned dry hole or a producer, may be called a completion.

Condensate — The stock tank liquid formed by condensation, produced from a reservoir that originally contained hydrocarbons in the gas phase. Lease liquid recovery from a gas well.

Conductor Pipe — A short string of casing of large diameter, the principal function being to keep the top of the well bore open and to provide means of conveying the upflowing drilling fluid from the well bore to the slush pit until surface casing is set.

Coning — Coning is an effect usually associated with high producing rates. In the case of water, the bottom water is attracted upward when the pressure differential over the producing interval is increased beyond the critical limits. In the case of gas, the gas is attracted downward by the same phenomena. A condition, detrimental to good vapor-liquid contact, existing when vapor velocity through bubble cap slots is excessive and gas pushes the liquid away from the slots.

Connate Water — Water that fills a portion of the pore

Glossary

space containing oil or gas at the time of discovery.

Conservation — The productin of unrenewable natural resources with maximum efficiency and minimum waste.

Contour Map — A map on which points of equal elevations are represented by a line called a contour. Contour map in the oil industry refers to graphical depiction of the topography of a subsurface formation.

Control Valve — An automatically actuated valve for controlling flow, in response to an impulse from an instrument which measures the flow, or some function affected by it.

Core Barrel — A device used in rotary drilling to cut cores. The core barrel, varying in length from 25 to 60 feet, is run at the bottom of the drill pipe in place of a bit, or in conjunction with a special type of bit.

Coring — The act of procuring large samples of the formation being drilled for informational purposes.

Correlation — The determination of the same geological point in two or more wells.

Corrosion — The destruction of a metal by chemical or electrochemical reaction with its environment.

Counterbalance — A weight attached to a crank, pulley rim, walking beam or other moving part of a pumping unit providing for even distribution of loads and for the reduction of peak torque during the up and down stroke.

Crater — A large sink hole or cavity around a well. Sometimes accompanies a violent blowout during which the surface surrounding the well drops.

Crawlers — Mechanical devices which are pumped through flow lines to remove unwanted deposits.

Crooked Hole — One that has numerous unintentional deviations from the vertical.

Cut Oil — Oil which is partly emulsified with water.

Cuttings — The rock or formation particles cut in the drilling process and brought to the surface by the drilling mud.

Glossary

Datum Horizon — A horizon used as a reference for elevations. For most topographic work, the datum plane is mean sea level. In structural mapping, the bed or horizon to which all elevations are finally reduced is called datum horizon, key bed, or market. For bottomhole pressure mapping, the datum is usually selected at the approximate mid-elevation of the reservoir referred to sea level.

Dead Man — A timber, concrete block, metal block or pipe buried in the earth to which a line may be attached, thus serving as an anchor.

Decline — The decrease in yield of oil or gas from a well, lease, pool or field. The first yield is called the flush production. For awhile, the decline is rapid, becoming more steady until settled production is reached. Decline curves in which yield is plotted against time show graphically the change in rate of production.

Degasser — Equipment used to remove undesired gas from a liquid, especially from drilling fluid.

Depletion — The act of emptying, reducing or exhausting, as the depletion of natural resources.

Depreciation — Reduction in capital value of a tangible asset (such as mechanical equipment, building, etc.) that results from wear, waste and obsolescence.

Desander — A centrifugal device for removing sand from drilling fluid to prevent abrasion of the pumps.

Dew Point Tester — An instrument for visually determining the temperature at which moisture condenses from a gas.

Differential Fill-Up Collar — A device used in setting casing. It is run near the bottom of the casing and automatically admits drilling fluid into the casing as required to cause the casing to sink rather than "float" in the well.

Differential Pressure — The difference in pressure between two points in a fluid system. It may represent the drop in pressure of the fluid in passing from the tubing to the flow line. In the case of an orifice meter, the difference of

Glossary

the pressure on the up-stream and the down-stream sides of the orifice.

Directional Drilling — Intentional drilling of an off-vertical well at a closely controlled, predetermined angle and direction through the use of special equipment and surveys.

Discounted R.O.I. (Return on Investment) — The ratio of the (1) present worth of the ultimate value of the reserves attributable to a well less the drilling investment, taxes, operating costs and other needed investment anticipated during the life of the well, to (2) the present worth of the drilling and other investments anticipated during the life of the well.

Distillate — The condensed hydrocarbons which are produced with gas from a single or two phase reservoir. Condensation occurs as pressure is reduced below a certain critical pressure. Also, the condensed overhead product of a fractionator.

Dog House — The small building where the drillers and roughnecks change clothes and which serves as an office for keeping records. Also a small building used as an office by the pumper.

Dog Leg — Term applied to a sharp change of direction in the well bore.

Dollie — A low frame with wheel or rollers designed to support heavy loads to be moved, as casing dollie to support the end of casing as it is taken into the derrick from the walkway.

Dope — A lubricant for the threads of oil field tubular goods.

Double — A section of drill pipe, casing or tubing consisting of two joints screwed together.

Drawdown — The difference between levels in a water well when the pump is not working (static level) and when it is working (pumping level).

Drawworks — The hoisting mechanism on a drilling rig.

Glossary

It is essentially a large winch which spools off or takes in the drilling line and thus raises or lowers the drill string and bit.

Drill Collar — Heavy, thick-walled tube, usually steel, employed between the drill pipe and the bit in the drill string to provide weight on the bit in order to improve its performance.

Drill Pipe (Drill Stem) — The pipe used in drilling a rotary well. In modern drilling methods, this pipe may be required to connect in a single string of 20,000 feet or more.

Drill Stem Test (DST) — A test through the drill pipe or tubing taken by means of special testing equipment to determine if oil or gas in commercial quantities has been encountered in the well bore. It is not necessary to set casing or to remove drilling fluid from the hole to make a drill stem test.

Drilling In — The process of drilling into or through the oil or gas pay for completion.

Drilling Out — This refers to drilling out of the residual cement which normally remains in the lower section of casing and the well bore after the casing has been cemented.

Drilling Slot — Opening in a drilling platform or drilling vessel through which drilling operations are conducted.

Drip — A vessel attached to a natural gas well or gas line to arrest and accumulate any liquid that might find its way into the gas mains.

Dry Gas — Natural gas obtained from sands that produce gas only. It contains a negligible quantitiy of the heavier fractions which may be recovered as liquid.

Dry Hole — In general, any well that does not produce oil or gas in commercial quantities.

Dump Bailer — A bailer equipped to dump cement, water, mud, etc. in the bottom of a well.

Dump Valve — A valve used to reduce liquid level in a vessel — usually an automatic drain valve.

Glossary

Dynamometer — As applied to a sucker rod load measurements, an instrument which measures the polished rod loads in relation to the position of the rod within the pumping cycle.

Edge Water — Formation water occupying the perimeters of an oil and gas reservoir.

Electric Logging — Measurement of resistivity and self potential of formations immediately adjacent to a well bore by passing electrodes down the well.

Electric Pilot — An instrument used in well surveying, which utilizes an electrical current in connection with two fluids, one a conductor and one a non-conductor of electricity to determine accurately the interface between the two fluids.

Elevator — A hinged circle or latch block provided with long links to hang on the elevator hook used to hoist or lower drill pipe, casing, tubing, and sucker rods.

Emulsion — A mixture of two immiscible liquids; that is, liquids which do not mix together under normal conditions.

Entrainment — Mist which is carried in a vapor or gas stream.

Equivalent Weight — The molecular weight of a substance divided by its hydrogen equivalent. For most compounds, the molecular weight divided by the total positive or negative valence.

Exchanger — Technically, any equipment used for the indirect transfer of heat. Conventionally, a specific piece of equipment used to conserve energy by transferring heat from a process stream being cooled to one being heated.

Expansion Joint — A device used in connecting up pipe lines to permit linear expansion or contraction as the temperature rises or falls. "U" shaped bends or coils also used for this purpose.

Extraction — The percent of a given component of the plant inlet gas which is removed from the gas stream by

Glossary

absorption or condensation.

Fault — A geological term denoting a break in the subsurface strata. Usually strata on one side of the fault line have been displaced upward or downward relative to their original position.

F.F. Flange — Flat Face Flange. A type of flange having a flat mating surface.

Filling the Hole — Pumping drilling fluid into the well bore as the pipe is being withdrawn in order to maintain the fluid level in the hole near the surface or to fill the well bore with any liquid.

Finger Board — A board located up in the derrick to support the upper end of stands of pipe standing on the derrick floor.

Fire Wall — Ordinarily a dike or earthen wall thrown up around a tank or a battery of tank to lessen the hazard of a fire spreading in case one develops.

Fish — A name given to any article lost in a well.

Fishing — An operation requiring special tools and a considerable degree of skill, whose purpose is the recovery of tools, casing or other articles lost or stuck in a well.

Fittings — A term used to denote all those pieces which may be attached to pipes in order to connect or provide outlets, etc.

Flange Union — A fitting consisting of a pair of threaded plates and bolts to connect threaded pipe.

Flare — An open flame used to dispose of unwanted gas.

Float Collar — A collar inserted one or two joints above the bottom of the casing string, and which contains a check valve that permits fluid to pass downward through the casing but prevents it from passing upward. The float collar prevents the drilling mud from entering the casing while it is being lowered, thus allowing the casing to "float" during its descent, and decreasing the load upon the derrick. The

Glossary

float collar also prevents the back flow of cement during the cementing operation.

Float Valve — A valve which is operated by a float. The valve in a float collar.

Flooding — The drowning of a well by water. Also a process by which oil is driven to the well bore by either designed water injection or natural water influx. The unstable operation of a packed column or bubble tower whereby the entire vessel or a section of it is full of liquid due to high liquid or gas loading. Pressure drop of the gas through the vessel under such conditions is excessive because of the high liquid head.

Flow Bean — A removable steel orifice or restriction that is placed in the flow line to restrict and control the rate of flow of gas and oil from a well. Sometimes called a choke or flow nipple.

Flow Chart — A systems analysis tool providing a graphical presentation of a procedure. Includes block diagrams, routine sequence diagrams, general flow charts, etc.

Flow Line — The pipe line connecting a well with a tank battery.

Flowing by Heads — Intermittent flowing.

Flowing Well — A well from which oil or water flows naturally without artificial aid.

Fluid Level — Ordinarily the distance from the surface to the top of liquid in the well bore.

Flush Production — The yield of an oil well during the early period of production and before the output has settled down to what may be regarded as usual for the field or district in which it is drilled.

Formation Damage — The reduction of permeability in a reservoir rock arising from the invasion of drilling fluid and treating fluids into the section adjacent to the well bore.

Formation Water — See "Connate Water"

Frac Job — A means of stimulating production by

Glossary

increasing the permeability of the producing formation. By use of a fluid, hydraulic pressure is applied to "fracture" the formation, causing cracks.

Free Gas — Gas occupying reservoir voids, but not in solution in oil.

Free Point Indicator — A tool designed to measure the amount of stretch in a string of stuck pipe, and in so doing to indicate the deepest point at which the pipe is free. The free point indicator is lowered into the well on a conducting cable.

Frost Box — A protective enclosure, usually wooden, erected around manifolds or hook-ups that handle water or water cut oil, as protection against freezing of the fluids.

Gall — To score or ridge a bearing or shaft.

Gas Cap — The portion of an oil reservoir filled with free gas from which the production is predominantly gas.

Gas Drive — The expansion of injected gas or free gas in a reservoir to force liquid hydrocarbons to the producing wells.

Gas Lock (Subsurface Oil Well Pumps) — Condition of gas in the pumping chamber being compressed and expanded between the valves with no fluid delivered.

Gas Lift — A method of mechanical lifting of oil in which the energy of compressed gas is used as the source of power for bringing the well fluid to the surface.

Gas-Oil Ratio (GOR) — **Gas Liquid Ratio** — The number of cubic feet of gas measured at a stated pressure base produced or associated with one barrel of oil (gas/oil ratio) or with one barrel of liquid (gas/liquid ratio).

Gate Valve — A valve used in piping to interrupt flow.

Gathering Lines — Pipes connecting tank batteries on oil leases with trunk pipe lines. In natural gas systems, the lines that connect with the wells and carry the gas to the main pipe lines.

Glossary

Gauge — To determine the quantity of liquid in a tank or vessel by reading the height of liquid.

Gauge Glass — A glass tube that furnishes a visual indication of the level of water or other liquid within a vessel.

Gauge Pressure — The pressure in pounds per square inch above atmosphere.

Gauger — An employee of the purchaser or transporter of oil, whose duty it is to take samples, test them for quality as to gravity and B.S. & W. content and to determine the net volumes sold or run to the pipe line.

Gin Pole — A pole used to support hoisting tackle. A V-shaped hoisting device normally located on the back of a truck.

Globe Valve — A valve used in piping to throttle and to interrupt flow.

Glycol — A liquid dehydrating agent, usually diethylene or triethylene glycol, used in the continuous drying of a natural gas.

Go-Devil — A scraper with self-adjusting spring blades, inserted in a pipe line and carried forward by the fluid pressure, clearing away accumulations, particularly paraffin, from the walls of the pipe. The term is also applied to a weight dropped to explode a charge of nitroglycerine placed at the bottom of the well. Also free falling devices designed to perform special functions such as measure inclination of the hole, release tools on the bottom of a wire line, etc.

Goose Neck — The connecting member of the hose on a rotary swivel.

GPM — Gallons per thousand. The gallons of liquifiable products in one thousand standard cubic feet of gas, also gallons per minute.

Grade — An elevation reference point employed in surveying.

Glossary

Gravel Pack — The practice of placing selected gravel opposite a producing formation to minimize its sloughing and filling up the well bore thereby reducing production. Gravel used as a filter — usually poured into a water well annulus outside of the screen.

Grind Out a Sample — A popular expression denoting the running of a centrifuge test on a sample of oil to determine the amount of B. S. & W.

Guide Shoe — A short, heavy cylindrical steel section filled with concrete and rounded at the bottom, which is placed at the bottom end of the casing string. The guide shoe prevents the casing from hanging up on irregularities in the bore hole as it is lowered.

Gun-Barrel — A wash or settling tank in which water settles out of oil.

Guy — A rope, chain, or rod attached to anything to steady it. A wire line attached to the top of a derrick and extending obliquely to the ground where it is fastened to a guy anchor.

Hardness (Water) — Undesirable scale-forming salts, principally calcium and magnesium.

Head (Of Liquid) — Pressure created by liquid due to the height at which the surface of the fluid stands above the point where the pressure is taken. Expressed in "pounds per square inch" or "feet." Also a single flow of a well when flowing intermittently.

Header — A pipe to which a number of lines, usually in the same service, are connected.

Hold Down Nipple — Usually a 1 foot portion of the tubing string in which the pump is seated. Commonly located above the gas anchor. Its function is to hold the stationary part of the pump in place during the upstroke.

Holiday — A hole, air pocket or break in pipe coating.

Hot Spot — An abnormally hot place on the tube in a boiler or furnace. Also a location of active corrosion of buried pipe.

Glossary

Hunting — The fluctuation observed in a control instrument when it is attempting to establish a stable operating condition. Also, uneven or cycling operation of an internal combustion engine due to bad timing or erractic fuel injection.

Hydrafrac — A copyrighted name of an operation whereby producing formations are fractured by applying hydraulic pressure to the formation through highly viscous fluids to increase productivity. The fluids sometimes carry sand to aid in preventing closure of the induced fractures.

Hydrate — A chemical compound of water and gaseous hydrocarbons, or hydrogen sulfide, normally formed at pressures exceeding 100 psi. Similar to porous snow in appearance, this solid has a specific gravity close to that of water. Accumulation of hydrates in pipe lines and processing equipment results in freeze-ups, a considerable source of trouble.

Hydrogen Sulfide — A gaseous compound of sulfur and hydrogen having the odor of rotten eggs, occurring frequently in crude oil, gas and water. Hydrogen sulfide is toxic.

Hydrostatic Pressure — The pressure exerted by a column of fluid.

Impervious — Impassable; applied to strata such as clays, shales, etc., which will not permit the passage of water, petroleum or natural gas.

Impression Block — A lead or wooden block which is run in a well and allowed to rest on top of lost tools or other fish to obtain an impression of the pin or top of the fish lost in the hole for use in planning fishing operations.

Injectivity Test — A test made on a well to determine the rate at which the well will take an injected medium at a given pressure.

Integrator (Chart) — A mechanical device employed to calculate the cumulative or total flow recorded on a flow meter chart.

Glossary

Intermediate String — The middle string of casing where three strings of casing are run in a well. Sometimes called protection string.

"J" Slot — A locking arrangement utilizing a slot in the form of a "J" and a pin which prevents premature setting of bottomhole equipment and at the same time provides a method of surface control for setting the equipment at the proper time and depth.

Jack-screw — A screw or bolt used to spread two elements apart.

Jack-shaft — A short shaft between the couplings on a rotating unit and its drive.

Jars — A drilling tool having a sliding interlocking joint which affords flexible motion.

Jet Perforating — An operation similar to gun perforating except that a shaped charge of high explosives is used to burn a hole through the casing instead of using the gun which fires a projectile.

Joints — The place or part where two lengths of tubular goods or sucker rods are joined or united. Oil field vernacular usually refers to one length of tubular goods as one joint.

Kelly — The heavy square or hexagonal steel member which is suspended from the swivel through the rotary table and connected to the drill pipe to turn the drill string.

Kelly Bushing — Device fitted to the rotary table through which the kelly passes and by means of which the torque of the rotary table is transmitted to the kelly and to the drill stem. Sometimes called drive bushing.

Kelly Cock — A valve installed between the swivel and kelly. When a high pressure backflow begins, the operator can close this valve and keep the pressure off the swivel and rotary hose.

Key Seat — A channel or groove cut in the side of the bore

Glossary

hole, parallel to the axis of the hole, on the inside curve of a dog leg.

Killing a Well — The act of bringing under control a well which is blowing out; also applied to the procedure of circulating water and mud into a completed well before starting workover operations.

LACT Unit — (Abbreviation for *Lease Automatic Custody Transfer*.) An equipment assembly designed to measure, sample, and record crude oil volumes and properties while delivering oil to the pipe lines on an unattended basis.

Land Casing — To set the casing at a given depth in the hole.

Latch-on — To attach elevators to a section of pipe.

Laterals — The pipe lines which are connected to a header.

Lead Acetate Test — A test for determining whether the hydrogen sulfide content of a gas stream is in excess of about 0.25 grains per 100 cubic feet.

Lessee — The one to whom a lease is granted.

Lessor — The one who grants the lease.

Lifting Cost — The expense (operating) of producing oil, ordinarily referred to on a per barrel or per well month basis.

Line Out — To adjust the necessary variables in a particular process unit to obtain satisfactory operation.

Liner — A short string of casing, set at the bottom of the hole and extending into the next larger string of casing. Also the lining of barrels of pumps.

LP-Gas (Liquefied Petroleum Gas) — The hydrocarbon products commonly known as butane, propane and mixtures of these two. They occur between what is commonly known as gasoline and dry natural gas in the hydrocarbon series.

Load Factor — The ratio of average load to the maximum demand during a given period, expressed as a percentage.

Glossary

Load the Hole — The action of filling the well with some liquid to provide a head to control high pressure formations and to prevent acid, Hydrafrac process liquids, etc. from being displaced up the hole instead of into formation which is being treated.

Location — A spot or place where a well is to be drilled.

Log — A detailed record of the nature of the formations penetrated during drilling. Data recorded may consist of electrical and radioactive surveys, description of cuttings, core analyses, etc. correlated with depths. Also refers to a history of operations where drilling time, intervals cored, drill stem test results, etc. are recorded.

Loop — A section of pipe substantially parallel with, and connected at both ends to another line to increase flow rate or to decrease pressure drop.

Lost Returns (Lost Circulation) — Loss of drilling fluid into a formation in the well bore during drilling operations.

Lubricator — In the oil field, in addition to its ordinary meaning, the term, is applied to special devices for introducing chemicals, fluids, and instruments into a system under pressure. A lubricator for a bottomhole pressure bomb would consist of a piece of tubing long enough to hole the instrument, screwed into a valve on the wellhead with a stuffing box on top of the piece of tubing which would permit introducing the bomb to the well without releasing the pressure.

Make Tank — A tank in which a product ("Make") is stored.

Make-up — To screw together, as drill pipe or a string of tools.

M-alkalinity Test — The method for determining the alkalinity present in water up to the methyl orange end point.

Manifold — A pipe header with numerous branches or outlets.

Glossary

Marginal Well — A low capacity well or one having a small productivity and approaching its economic limit of operation. Generally, used in a similar manner as stripper well. The term is specifically defined in statutes of some states and may vary from state to state.

Marsh Funnel — A calibrated funnel commonly used in field tests to determine the viscosity of drilling mud.

Mast — A long pole or tube rising in an almost vertical direction to support some member of a drilling or hoisting machine.

Material Balance — A mass balance in which all materials entering an area under study are equated to all materials leaving the area. Used as a tool in determining quantity of an unknown stream, verifying metered streams, etc.

MEA — Monoethanolamine. An organic base used in an amine unit to remove hydrogen sulfide and carbon dioxide from a gas stream.

MER — Has two general meanings, the first of which is the most common: (1) "Maximum efficient rate" is the highest rate at which a well or reservoir may be produced without causing physical waste in the reservoir. In practice, economics are generally considered and an MER below the minimum rate at which a well or reservoir can be economically produced is not considered. (2) "Most efficient rate" is the highest rate at which a reservoir can be produced without either reservoir or surface physical waste. For example, a reservoir may be produced at the "Maximum efficient rate", but at such a rate, gas production will be in excess of the capacity of facilities in the field to handle the gas, so a lower or "most efficient rate" is set up for the reservoir to avoid surface waste of valuable hydrocarbons in the form of flared gas.

Meter Run — Pipe on each side of an orifice plate in an orifice meter installation of sufficient length to permit uniform flow and representative measurements of the static and differential pressures.

Glossary

Migration — The movement of oil or gas through the pores of the rock, especially with reference to movement across lease boundaries or from one portion of the reservoir to another. The rate of movement varies with the permeability of the rock, the viscosity of oil, existing pressure gradients and other factors.

Mill — A tool having a rough, sharp and extremely hard cutting head for removing metal by grinding or cutting. Mills are run on drill pipe or tubing and are used for such purposes as grinding up debris in the hole, removing sections of casing for sidetracking and reaming out tight spots in the casing. They are variously termed junk mills, reaming mills, etc., according to the use for which they were designed.

Mist Extractor — A device to remove entrained liquid from a gas stream. May be installed inside of a vessel or in the piping outside a vessel.

Mouse Hole — A hole drilled under the derrick floor and temporarily cased in which a length of drill pipe is temporarily suspended for later connection to the drill string.

MSCFD — One thousand cubic feet of gas per day measured at 60°F and 14.7 psia.

Mud — A mixture of clay or earth minerals with water. Used as a drilling fluid.

Mud Cake — The sheath of mud solids which forms on the wall of the well when the liquid from the mud filters into the formation.

Mud Logging — The recording of information derived from examination and analysis of return circulation mud and drill bit cuttings. A portion of the mud is diverted through a gas detecting device and examined further under ultraviolet light for the purpose of detecting the presence of oil or gas.

Mud Pump — A pump used to pump mud in order to maintain circulation in a drilling well.

Glossary

Mud Screen — A shale shaker. The vibrating screen that is used to remove cuttings from the mud as it returns to the surface from the bottom of the well.

Mud Weight — The weight of a gallon of drilling mud.

Natural Gas — A hydrocarbon gas of variable composition found in nature, which usually has a high methane content.

Natural Flow — Production of oil at the surface by use of available reservoir energy which may be water pressure, free gas pressure, solution gas or a combination of these factors and without the use of artificial lift methods.

Nipple — A short pipe threaded on both ends and used to connect up fittings in a convenient manner.

Normal BHP — A reservoir is said to have a normal BHP when the initial BHP approximates the hydrostatic pressure of a column of salt water corresponding to the depth of the reservoir.

Nozzle — A flanged piping connection on a vessel, pump, etc.

Nutating Disc — Wobble disc used in a liquid flowmeter to transfer motion of the flowing fluid to the counter.

Offset Well — A well drilled opposite a well on an adjoining property, the distance from the common boundary depending on the well spacing.

Oil Column or Water Column — Thickness of an oil or water layer in the reservoir rock.

Oil Reserve (An Estimate) — Oil remaining underground, reasonably proved productive, the recovery of which is commercially feasible.

Oil Saver — A device used to prevent the escape of oil and gas from the well when pulling or running tubular goods, sucker rods, or swab line. Rubbers are normally used as the wiping and sealing medium.

Oil String — A string of casing used to protect and to keep

Glossary

the oil well open through the rock formations, down to or through the producing formation.

Oil Squeeze — The forcing of oil into a formation to improve the producing characteristics of a well.

Open Ended — A piece of pipe or tubing not equipped with a fitting for closing it.

Open Flow Test — A series of tests made from which the volume of gas that will flow from a well in a given time when the well is open to the atmosphere may be calculated.

Open Hole — The uncased part of the well.

Orifice Meter — A gas or liquid flow measuring device employing a thin plate with an orifice inserted in a pipe line to create a pressure differential and having a mechanism for indicating or recording the amount of pressure differential. When measuring gas, due to its compressibility, the static pressure of the gas on the line must also be determined.

Outage — The difference between the full interior volume of a storage vessel and the volume of liquid therein. For gasoline and lighter products, the regulatory bodies set a minimum limit for outage in order to provide space for expansion of the liquid.

Overburden Pressure — The pressure at a given depth in the earth's surface exerted by the overlying formations.

Overshot — A fishing tool used to lower over the lost or stuck pipe or sucker rods thereby obtaining a frictional grip permitting recovery. It is the female counterpart of a spear.

Packer — A device, usually employing rubber, used to effect a seal between tubing, casing, or drill pipe and the open hole or casing.

P-alkalinity Test — The method for determining the alkalinity present in water at the phenolphthalein end point.

Paraffin — A waxy-like hydrocarbon of varying hardness

Glossary

and melting point which precipitates out of oil and deposits on the producing formation and tubular goods hindering normal producing operations.

Pass — One of several similar repetitive operations (e.g., in pipe welding, a pass is one welding circuit of the pipe; in a tubular exchanger, a pass is the flow of fluid from one end to the other, either in the tubes or the shell).

Payout — Time, in months, required to payout the drilling investment (includes lease equipment chargeable to the well and pumping equipment if needed).

Perforate — Pierce with holes as is done to well casing after it has been set in the well to allow flow of the reservoir fluids to the surface.

Permeable — A property of rock denoting its ability to pass fluids and commonly used by members of the oil industry to distinguish between rocks which will give up no fluids and those which will produce oil, gas and water. The measure of permeability, millidarcies, is a means of direct comparison of one rock's fluid transporting ability to that of another's. Some rocks may be porous but not permeable.

Pipe Line Connection — The outlet from a well or tank by which gas or oil is transferred to a pipe line for transportation away from the field.

Pipe Line Oil — Oil of sufficient purity to meet the specifications of the purchaser.

Platform — Structure built for use at water locations to support equipment for drilling or producing operations.

Plug — A pipe fitting or piece of suitable material designed to fill a hole. In oil wells, it may be used to shut off gas or other undesirable fluids.

Plug Back — The act of partly filling a well with impervious materials for the purpose of shutting off lower formations in order to permit upper formations to be produced.

Polished Rod Clamp — A clamp which grips the polished rod of a pumping well.

Glossary

Polished Rod — A rod with polished surface at upper end of sucker rod string, which passes through the stuffing box of a pumping well.

Pop Valve — A pressure relieving safety valve.

Pore Space — The open space, or voids, between the individual grains of a rock mass, available for fluid accumulation.

Porosity — A property of rock denoting its percentage of pore space of the total volume of the rock mass.

Potential Test — The production, generally measured in barrels of oil or standard cubic feet of gas per 24 hour day, that a well will produce under certain specified conditions.

Pounding Fluid — The striking down on fluid by the traveling valve of a sucker rod pump due to pump barrel being only partially filled.

Present Worth — The principal of a sum of money payable at a future date, such that this principal plus all accrued interest at the given interest rate will amount to the sum at the date on which the sum is to be paid (also referred to as the discounted value of that sum which is to be paid).

Pressure Base — An absolute pressure expressed in pounds per square inch absolute (psia) agreed upon as a basis for comparison of volumes of gases measured at different pressures.

Pressure Gauge — A device which measures fluid or gas pressure, ordinarily reading in pounds per square inch above atmospheric pressure (psig).

Pressure Maintenance — A program to prevent or control decline of reservoir pressure as the reservoir fluids are produced, to increase ultimate recovery and profit, ordinarily accomplished by the injection of fluids such as gas and water.

Primary Production — That portion of a reservoir's reserves that is recovered due to the energy content of the reservoir and surrounding fluid, and without the injection of gas, water or other sources of energy.

Glossary

Production — A term used to indicate the phase of the petroleum industry that deals with the extraction of the oil and gas from the ground; the amount of oil produced, viz. daily production, monthly production, flush production, settles production, etc.

Production Curves — Curves plotted to show the producing characteristics of individual wells, leases, pools or fields, made particularly for the purpose of studying the reservoir and the rate of production decline of wells and pools.

Productivity Index (P.I.) — A measure of a well's ability to produce, expressed in terms of barrels of oil produced per day per pound of differential pressure between the static reservoir pressure and the well's flowing bottomhole pressure.

Profitability Index — That discount rate which equates the present worth of the cash income and outflow and is the average rate of return earned on the money while it is actually invested in the project.

Proppant — Granular material used to prop open hydraulically created fractures.

Prorate — As applied in the petroleum industry, pertains to the allocation of oil or gas production among the properties producing from a common reservoir or among the fields in a given state on some agreed or enforced basis. Among the factors used as a basis for proration are acreage, number of wells, well potential, or a combination of these and other factors.

Pulling Costs — Expense of servicing wells by pulling and repairing rods, tubing or pump.

Pulling Line — The cable on a servicing unit winch used to raise and lower the rods and tubing in the derrick while servicing a well.

Pulling Tool — A hydraulically operated tool that is run in above the fishing tool and anchored to the casing by slips. By means of hydraulic power derived from fluid that is

Glossary

pumped down the fishing string, the pulling tool exerts a strong upward pull on the fish.

Pumper — A workman who produces oil wells; also an oil producing well which does not flow its production and requires artificial lift.

Racking Pipe — The act of placing stands of drill pipe or tubing in orderly arrangement in the derrick while hoisting pipe from the well bore.

Rat Hole — A slanting hole located inside and near a corner of the derrick. It is cased with a large size length of casing and is about 25 feet deep. It forms a depository for the kelly so that the swivel, which connects to the upper end of the kelly, can be conveniently adjusted and lubricated. It also provides an out-of-the-way place for the kelly, swivel and kindred parts when drill pipe or casing is being run into the well.

Reaming — The operation of enlarging an existing well bore. Used to enlarge an undersized core hole so that the regular drilling bit can be run to bottom and drilling continued. See "Underream."

Recompletion — The completion of a well in another reservoir either by drilling deeper or plugging back.

Regulator — A piece of equipment used to control pressure.

Repressuring — A program of increasing reservoir pressure by injecting gas and water into one or more wells in the same horizon to increase oil producing rate and ultimate oil recovery.

Re-Run — To process an off-specificating material to produce an acceptable product.

Reservoir — The portion of a formation in which oil and gas has accumulated.

Return on Investment (R.O.I.) — Ratio of (1) ultimate value of the reserves attributable to a well less the drilling investment, taxes, operating costs, and other needed

Glossary

investments anticipated during the life of a well, to (2) the drilling and other investments anticipated during the life of the well.

Reservoir Energy — The energy inherent in an oil or gas reservoir which forces the reservoir fluid to the well bores. In oil reservoirs, this energy may be from one or more sources — (1) expansion of compressed oil, water or gas, (2) evolution of dissolved gas, (3) gravity forces, or (4) water encroachment. In gas reservoirs, the energy is usually (1) expansion of gas or water, or (2) water encroachment.

Retention — The ratio of recovery to extraction for a given component, expressed as a percentage.

Retrievable — Ordinarily has reference to special subsurface equipment which after use can be released and recovered from the well bore.

Reverse Circulation — Circulating liquid returns to the surface through the drill pipe after being pumped down the annular space.

R.F. Flange — Raised Face Flange. A type of flange having the gasket contact surface raised.

Rigging Up — The work of installing a boiler, engine, tools and machinery and establishing a supply of fuel and water prior to start of drilling.

Rock Pressure — Usually applies to the actual formation pressure of a gas reservoir and is equal to the wellhead pressure plus the weight of the gas column.

Rocking the Well — The pressure required to start a well flowing by gas lift is two to five times that needed when the well is in normal operations. This is because the column of oil in the well, being unsaturated with gas, is much more dense than the mixture of oil and gas that flows after regular production has been established. To keep this starting pressure at as low a point as possible to avoid forcing oil back into the formation, the casinghead sometimes is equipped to reverse the flow of gas down into the well. Thus, it is possible to force gas first down the

Glossary

tubing and then down between the tubing and casing. Several of these reversals of the gas are required sometimes to get the well started. This manipulation of the gas flow is called rocking the well because of its back-and-forth character.

Rod Guide — An attachment to sucker rods which serves to prevent them from rubbing on the sides of the tubing.

Rotary Bushing — A metal lining that fits into the rotary table opening to reduce its size for special purposes such as to fit the slips or to fit the kelly bushing. Usually called master bushing.

Rotary Drilling — Penetration of the earth by rotating a bit at the bottom of a hole.

Rotary Hose — On a drilling rig, the hose that conducts the drilling fluid from the mud pump and standpipe up to the swivel and kelly. Also termed mud hose.

Roughneck — A driller's helper and general all around worker on a drilling rig.

Round Trip — Pulling the drill pipe, usually for changing the bit, or tubing and rerunning to the bottom of the hole.

Roustabout — An employee who does general oil field work related to the production, transportation, treating and storing of oil and gas.

RTJ — Ring Type Joint. A type of flanged joint with grooved flanges and a loose ring.

Run — Oil taken from a tank by a pipe line.

Saddle — A fitting made in parts to clamp onto a pipe for the purpose of stopping a leak or providing an outlet. A welded fitting used to reinforce an opening in a pipe line.

Safety Head — A pressure relieving safety device, containing a frangible disc designed to break when pressure on one side exceeds a specified amount.

Safety Joint — A fishing tool accessory placed above the tool. If the tool is engaged and the fish cannot be pulled, the

Glossary

safety joint will permit disengagement.

Safety Latch — A latch provided in a hook or elevator to prevent these devices from opening and dropping the weight prematurely.

Safety Valve — An automatic valve used for the release of pressure when a certain limiting pressure is exceeded. Also called Pop Valve.

Sand Pump — A cylinder with a plunger inside and a valve at the bottom, lowered into a drill hole from time to time to take out the accumulated slime resulting from the action of the drill on the rock. Called also shell pump and sludger. Also, a pump for artificially lifting wells producing fluids containing sand.

Sanding-Up — The accumulation of sand at the bottom of a well resulting in reduced production.

Scale — A deposit precipitated out of water onto surfaces contacting the water.

Scratcher — A device fastened to casing which removes the mud cake from the wall of the hole to condition it for cementing. It is fashioned of stiff wire.

Screen — A special perforated pipe or casing which has the perforations protected by screen, usually used in loosely consolidated sand formations.

Scrubber — A vessel in which liquids are removed from a mixture of gas and liquids.

Section Milling — Descriptive term of the process by which a portion of pipe — casing — is actually removed by a cutting operation involving a mill.

Separator — A closed steel vessel or tank of special design having interior baffles and automatic regulating valves used to separate gas from oil as they flow from a well.

Series — Nominal working pressure of flanges and flanged valves at a specific temperature (ASA code).

Set Casing — See "Land Casing"

Settled Production — The production of an oil well or a

Glossary

pool during the period when the decline rate is rather slow as distinguished from the period of flush production.

Shot Feeder — A small pot through which liquid (usually chemicals) can be injected into a system under pressure.

Show — An indication of oil or gas while drilling a well.

Shrinkage — The reduction in volume of a gas stream due to removal of hydrocarbon products, hydrogen sulfide, or carbon dioxide. Also, the unaccounted loss of products from storage tanks. Also, loss in crude volume from reservoir when gas is evolved from solution.

Shut-In Tubing Pressure — The pressure noted at the wellhead when the well is completely shut-in. Tubing and casing pressures are usually different due to different heights of liquid columns.

Shut-Off — The successful exclusion of any undesirable fluid from the well bore through the use of packers, cement, mud, plastic, etc.

Side Tracking — Drilling past a broken drill, casing or other equipment which has become permanently lodged in the hole and cannot economically be removed or drilled up. See "Directional Drilling."

Sidewall Coring — A coring technique in which core samples are obtained from a zone that has already been drilled. In this technique, a hollow bullet is fired into the formation wall, capturing the core. It is retrieved on a flexible steel cable. Core samples of this type normally range from 3/4 to 1 3/16 inch in diameter and 3/4 to 1 inch in length.

Single — One joint of drill pipe.

Skid — A prefabricated base for an equipment assembly.

Skimming Pit — A pit into which oil and water is placed, the oil later being skimmed from the surface.

Slacking Off — Releasing the tension either on casing, drill pipe or tubing at the wellhead.

Slips — Wedge-shaped pieces of metal with teeth or other

Glossary

gripping elements used to prevent pipe from slipping down into the hole or for otherwise holding pipe in place. Rotary slips fit around the drill pipe and wedge against the master bushing to support the pipe. Power slips are pneumatically or hydraulically actuated devices operated by the driller at this station and which dispense with the manual handling of slips when making a connection. Packers and other down-hole equipment are secured in position by means of slips that are caused to engage the pipe by action performed at the surface.

Slip Tube — A device for gauging tank cars, trucks or storage tanks.

Sloughing — Parting of the formation during drilling operations, allowing particles of the formation to fall into the bore hole.

Slug — Any relatively large mass of liquid in a normally dry gas stream.

Slug the Pipe — Before hoisting drill pipe, it is desirable to pump into the top section of it a quantity of heavy mud — a slug — which will cause the level of the fluid in the pipe to fall. Thus, when a stand of pipe is unscrewed, the drilling fluid will have been evacuated from it. This prevents crew members and the rig floor from becoming covered with the drilling fluid.

Slurry — The liquid state of the cementing material after water has been added. After it has set to a solid, it is called cement.

Slush Pit — The pit by a drilling well into which the mud and cuttings are discharged. Also called slush pond.

Solution Gas-Oil Ratio — The cubic feet of gas initially dissolved in a barrel of oil in the reservoir.

Sonde — A logging tool assembly.

Sour Gas — Gas containing an appreciable quantity of hydrogen sulfide.

Sour Liquid — A liquid containing mercaptans or hydrogen sulfide. The term "doctor sour" indicates mercaptan contamination.

Glossary

Sour Oil — Oil containing an objectionable amount of sulfur or sulfur compounds.

Spacing — The number of acres per well, or the distance between wells.

Spear — A fishing tool which goes inside pipe lost in a well, in order to obtain a friction grip and permit recovery. It is the male counterpart of an overshot.

Spider — A solid steel casting designed to hold slips for the purpose of gripping and holding the tubing or casing while connecting another joint to the string. A device to hold a suspended string of casing in a well.

Spinning Line — A rope or chain coiled around a joint of casing or pipe and drawn around the cathead used in making up or unscrewing a joint of drill pipe or casing during the period that the pipe turns easily. In unscrewing by the use of tongs; also in finishing the making up of a joint, tongs are applied.

Spool — A short section of pipe with flanged ends.

"Spot" Oil — To place oil at a selected depth in a well in order to lubricate stuck pipe, clean the formation, or to prepare for acidizing, Hydrafrac treating, etc.

Spotting — The technique of placing a quantity of fluid — oil, water, acid, cement, etc. — at a desired position in the well bore.

Spudding In — The very beginning of drilling operations of a well.

Squeeze Job — Usually a secondary cementing job where cement is pumped into the formation through the bottom of the casing or through perforations to obtain a shut-off of undesirable fluids.

Stabbing — The act of guiding the end of a joint of pipe as it enters the coupling of another joint.

Stabbing Board — A temporary platform erected in the derrick at an elevation of 20 to 40 feet above the derrick floor. The derrickman or other crew member works on this board while casing is being run in a well.

Glossary

Stabilizer — A tool placed near the bit in the drilling assembly to change the deviation angle in a well by controlling the location of the contact point between the hole and drill collars.

Stand — Two or more joints of pipe, either tubing or drill stem, screwed together.

Standard Conditions — A combination of pressure and temperature used as a base for comparison of gas or vapor quantities. In engineering practice, a temperature of 60°F and a pressure of 14.7 pounds per square inch absolute.

Stand-Off Shooting Distance (Perforating) — The distance from the perforating gun to the interior casing surface.

Standing Valve — A stationary valve at the lower end of the working barrel of a sucker rod well pump.

Standpipe — A rig pipe, which is a part of the drilling fluid circulating system, extending up into the derrick to a height suitable for attaching the rotary hose.

Static Pressure — The pressure exerted at any given point by the weight of fluid above that point plus any pressure to which the fluid is subjected.

Stiff Leg — A type of derrick or crane.

Stop-Cocking — The practice of alternately closing and opening a stop-cock placed in the tubing near the connection to the flow line to release accumulations of oil and gas under pressure in a well.

Straddle Packer — As the term might indicate, a packer set above and below a given formation thereby isolating it from all other formations encountered in the well bore.

Strap — To calibrate a tank or measure the drill pipe while pulling out of the hole.

String Shot — An explosive method to back off stuck pipe utilizing primacord.

Stripper — See "Still"

Stroke — The vertical distance the sucker rods travel in a

Glossary

pumping well, as measured at the polished rod.

Stuffing Box — A part of a pumpiing well wellhead through which the polished rod passes. Also the chamber designed to contain packing and to maintain a fluid tight joint about a piston rod.

Sub — Short threaded pieces used to adapt parts of the drilling string which cannot otherwise be screwed together because of difference in thread size or design.

Substructure — The foundation on which the derrick and engines sit. Contains space for storage and well control equipment.

Sucker Rod — A metallic rod with screw connection at the ends providing a means for connecting with other rods forming a series of string of rods used to extend down in a well to the working parts of the pump and to actuate same.

Sucker Rod Hanger, Sucker Rod Jack — A device used in the upper part of a derrick from which to suspend stands of sucker rods when they are pulled from the well.

Surface Pipe — The first string of casing set in a well. On some wells, it is necessary to set a temporary conductor pipe which should not be confused with surface pipe.

Swabbing — Operation of a swab (a plunger with flexible rubber cups and sleeves) on a line to bring well fluids to the surface. This is a temporary operation to determine whether or not the well can be made to flow or to remove undesirable liquids from the hole.

Swage — A short piece of pipe with one end smaller than the other.

Sweet Crude — Crude oil containing little or no sulfur. See "Sour Oil".

Swivel — A rotary tool which is hung from the rotary hook and traveling block. Its functions are (1) to suspend and permit free rotation of the kelly and drill string, and (2) to provide a connection for the rotary hose and a passageway for the flow of drilling fluid into the kelly and drill string.

Glossary

Tail Pipe — Pipe run in a well below a packer.

Tally — A record of the tubing or casing installed in a well. It reveals the length of each joint, the number of joints, and the overall length of the string after making allowances for thread makeup.

Tank Battery — A series of lease tanks and related equipment close together which are operated by means of common connections.

Tank Bottoms — The accumulation of water and settlings in a tank below the pipe line connection.

Tank Gauge — A visual method for determining the height of liquid in a tank.

Tap — To make a small connection to a vessel or to an existing pipeline.

Temperature Survey — An operation to determine temperatures at various depths in the well bore. This survey is used in instances where there is doubt as to proper cementing of the casing, to find the location of inflows of water or gas into the well bore, and for other reasons.

Thief — A small cylindrical vessel designed to take a sample from any depth in a tank.

Thief Hatch — Gauge Hatch — An opening provided with a hinged covering on the top of a tank for the pumper or gauger to use when he must thief a sample of the liquid content, or gauge the tank.

Tribble — A stand of three joints of pipe.

Throttle — To reduce the rate of flow of a fluid stream by partially closing a valve.

Tie-Down — An anchor to prevent movement of equipment.

Tongs — The large wrenches used for turning to make up or break out drill pipe, casing, tubing and other pipe; variously called casing tongs, rotary tongs, etc. according to the use for which they are designed. Power tongs are pneumatically or hydraulically operated tools that serve to spin the pipe up tight, and in some instances, to apply final makeup torque.

Tool Joint — A heavy, special alloy steel coupling element for drill pipe. Tool joints have coarse, tapered threads and seating shoulders designed to sustain the weight of the drill stem, to withstand the strain of frequent coupling and uncoupling and to provide a leakproof seal. The male section of the joint — the pin — is attached to one end of a length of drill pipe, and the female section — the box — is attached to the other end.

Total Depth (T.D.) — The greatest depth reached by a well bore.

Traveling Block — The block containing sheaves and provided with clevis and hook which is connected with the load hoisted or lowered in a derrick.

Tubing Anchor — A device run in a pumping well as an integral part of the tubing which employs friction or slips between the anchor and the casing to prevent tubing movement with respect to the casing.

Tubing Catcher — A device with slips provided to engage the walls of the casing and catch tubing in case same is dropped in a well.

Tubing Hanger — A device included in the wellhead hook-up and contained in the tubing head which, by use of a mandrel or slips, suspends and holds the tubing string.

Twist Off — To break the drill pipe in the hole, usually by torsional stress.

Underream — To enlarge a bore hole below the casing.

Unload the Hole — To remove mud, water or other fluid from the well bore to induce production.

Valve — A device by which the flow of a liquid or a gas may be regulated or controlled. A movable part within the valve body that opens or obstructs passage. The most common types used in connection with oil and gas production are the

Glossary

gate valve, globe valve, needle valve, angle valve and check valve.

V-Door — An opening in a side of a derrick at the floor level having the form of an inverted V. This opening is opposite the drawworks. It is used as an entry to bring in drill pipe and casing from the pipe rack.

Viscosity — The property of a fluid that determines its rate of flow and is closely related to its internal friction; it is a measure of the degree of its fluidity. The measure of fluidity is usually expressed as the time in seconds required for a given volume at 60°F to flow through a given aperture.

Waiting on Cement (W.O.C.) — After the casing has been cemented, it is necessary to suspend operations and allow time for the cement to set or harden in the well bore. The time during which operations are suspended is designated as waiting on cement.

Walking Beam — An oscillating bar or beam, pivoted at the center, used to actuate a sucker rod pump. In cable tool drilling, the walking beam transmits motion to the drilling tools.

Washover — A procedure wherein pipe, commonly called wash pipe, larger than the fish is slipped over the fish and usually rotated to bottom with circulation to free the fish.

Water Drive — The process of driving oil and gas by means of water to a well or wells. There are two principal kinds of water drives, one which embodies natural water pressure, and one which results from injecting water through a well or wells.

Water Knockout — The device for removing free water from the produced fluid.

Water-Table — A level below which the pores and crevices of a formation are filled with water.

Wellhead Assembly — A term applied to the assembly of the casinghead and the tubing head.

Well Servicing Unit (Pulling Unit) — Equipment,

Glossary

usually portable, used to pull and run rods, tubing and other equipment in wells.

Well Surveying — The act of running an instrument into a well to obtain some specific type of information.

Wet Gas — Natural gas that contains significant amounts of condensible hydrocarbon fractions.

Wet String — Refers to a string of tubing from which the standing valve has not been removed so that the pulling and unscrewing of each stand of tubing releases oil or water on to the derrick floor. May also apply to drill pipe.

Whipstock — A round steel shaft designed to set in a well at some predetermined depth for the purpose of deflecting the drilling tools; used in side tracking and directional drilling.

Wildcat — A well drilled in territory not proved productive of oil or gas.

Wild Well — An oil or gas well flowing out of control.

Working Barrel — The body of a pump used in oil wells. A cylinder containing the working parts of a well pump.

Workover — Any major work performed on a completed well such as shooting, acidizing, recompleting, plugging back, squeeze cementing, repairing casing, Hydrafrac treating, cleaning out, deepening, etc.

INDEX

abnormal bottomhole pressure, 112, *137; see also* bottomhole pressure
absorber, 145
accelerator, *137*
access roads, *see* roads
acid fracturing, *137*
acid gas, *137*
acidizing, *137*, 173, 179
acid procedures, 102, 103; *see also* frac job
adjacent wells, 133
affidavits, false, 113; assumed name, 68, 128
Alabama, 56
allowable, *137*
American Petroleum Institute (A.P.I.), *138*
amine unit, *137*
anchor, *138*, 177
angle valve, 178
annulus, *138*
Arkansas, 33
artificial lift, *138*
audit of run tickets; *see* run tickets
Austin Chalk formation, 17, 21
Authorization for Expenditures (A.F.E.), 15, 47, 83-90, 111-112 132, *137*; forms, 83; overcharges on, 83; pitfalls of, 83
Authorization for Retirement (A.F.R.), *137*

back-off a joint, *138*
back pressure, *138*
backup, *138*
backup tongs, 138
backup wrench, 138
backwashing, 138
bailer, *138*
ball and seat, *138*
bankers, 8, 68-69, 125, 135
barite, *139*
Barnett Shale formation, 17
barrel, *139*
Basic Sediment and Water (B.S.& W.), 93, *141*, 144, 154-155
basket sub, *139*
beam hanger, *139*
bearings, 119
bell nipple, *139*
bentonite, *139*
Big Saline formation, 17
bit, *139; see also* drill bit
blanket gas, *139*
bleeder, 139
blenders, 102
blind flange, *139*
blind rams, *139*
block, *139*
blowdown, *140*, 144
blowout, 35, 106, 112-113, 116-120, 129, 131, *140*, 146; control of, 117-118, 120; prevention of, 120
blowout preventers, 111-112, 114 119, 139, *140*
bonus, 8, 45, 93; per-acre, 3, 6
boomer, *140*
bore-hole, *140*
bottom, 140
bottomhole pressure (B.H.P.), *140*, 159, 162, 166

Index

bottomhole pressure bomb, *140*
bottom settlings (B.S.), *140*
Bradenhead, *140*; see also casinghead
break-out, *140*
breaking down, *141*
breather, *141*
breathing, *141*
bridge, *141*
bridge plug, *141*
bubble tower, 152
bubble tray, 144
bull plug, *141*
buried rigs, 113
burn pit, *141*
bushing, *141*
butane, 158
butterfly valve, *141*
bypass, *141*

Caddo formation, 17
cage, *142*
caliche, *142*
caliper loggings, *142*
cap, *142*
capital, working, 8
capping, *142*
cap rock, *142*
cash flow, 43, 51
casing, 48, 59-60, 62-63, 77, 79, 88, 100, 102-104, 106, 124, 127, 138, 140-141, *142*, 143-145, 147-151, 157-158, 161-163, 167, 169-171, 173-179
casing centralizer, *142*
casing charts, 104; see also well logs
casing clamps, *142*
casing connection, 79
casing dollie, 148
casinghead, 140, 168, 178
casinghead gas, *143*
casing joints, 143
casing point, 127
casing pressure, 103, *142*
casing shoe, *142*
casing string, 155

casing tongs, 176
catcher, 89
cathead, *143*, 173
catline, 143
cattle-guards, 96, 98
catwalk, *143*
caustic embrittlement, *143*
caustic test, *143*
caustic unit, *143*
cavity, salt, *143*
cellar, *143*
cement bond, 77
cementing, 143
centrifuge, *144*
chain tongs, *144*
channeling, *144*
chart changer, 114
cheater, *144*
check valve, *144*, 151, 178
chloride test, *144*
choke, 117, *144*, 152
christmas tree, 4, 30, 36, 88, 117, *144*
circulate, *144*
clay, 156, 161
cleaning out, 179
cleaning up locations; see locations
cleanout plate, *145*
clevis and hook, 177
closing mud pits; see mud pits
cost and wrap, *145*
collar, 77, 79, *145*, 151
collar locator, 75-76, 79
collar log, see well logs
column, *145*
competence, investigating for, 108, see also incompetence
completion, of wells, 10, 17, 22, 32, 35, 43, 45, 47, 53, 56, 108, 131, 134, *145*; procedures of, 57
completion consultant; see consultants
completion point, 127
completion rigs, 72, 90, 105, 109, 124
compression expense, 50

—*182*—

Index

compressor, 50-51
consultants, 90-91, 103-106;
   incompetent, 104
consulting firms, 91
condensate, *145*
conductor pipe, *145*
coning, *145*
connate water, *145*
conservation, *146*
contour map, *146*
contractors, 15, 86-87, 96, 113, 115; dirt, 15-16, 32, 86, 96-98; dirt, 32-33, 86; drilling, 16-17, 74, 115-116, 120, 129, 135; errors of, 95; trucking, 16
control valve, *146*
core barrel, 88-89, *146*; cutting head of, 89
core data, 89
core lab technicians, 88-89
core samples, 171
coring, 88-89, *146*, 171; procedure, 88
Corpus Christi, Texas, 75-76
correlation, *146*
corrosion, *146*
cost estimate, 34
cost overrun, 34-35, 133
Cotton Valley formation, 17
counterbalance, *146*
crater, *146*
crawlers, *146*
crooked hole, *146*
crown block, 139
crude oil, 37, 55, 58, 140, 158; price of, 58, 91
cut oil, *146*
cuttings, *146*; *see also* drill bit

daily supervision; *see* consultants
damages, 97-98, 111, 116-118, 120; lawsuits from, 113, 118
data sheet, 51
datum horizon, *147*
dead man, *147*
decision-making, 111

decline, *147*
deductions, tax, 124
deepening, 179
deep wells, 34, 63
degasser, *147*
dehydrator, 19
Department of Transportation (D.O.T.), 54
depletion, 47, 51, *147*; tax provisions for, 123
depreciation, *147*
derrick, 114, 129, 138-139, 143, 148, 151, 155, 161, 166-167, 173-175, 177-179
desander, *147*
dew point tester, *147*
differential fill-up collar, *147*
differential pressure, *147*
digital readout monitors, 103
directional drilling, 99, 120, *148*, 179
dirt contractors; *see* contractors
dirt equipment, 97
dirt work, 32
disasters, 111, 113
discounted return on investment, *148*
distillate, *148*
division order department; *see* oil companies
dog house, 115, *148*
dog leg, *148*, 158
dollie, *148*
dope, *148*
double, *148*
Dowell, 108
down-hole pump, 49
draining procedures, 50
drawdown, *148*
drill, 170, 171
drill bit, 61-64, 88, 119, 146, 149, 174; cutting action of, 114; cuttings from, 88, 161, 172
drill collar, 64, *149*, 174; *see also* collar
drilling, 3-4, 10, 22, 32, 35, 43, 46-48, 54, 60-62, 64, 77, 86-91, 97,

—*183*—

Index

112-114, 118-120, 123, 125-127, 131, 140, 143, 145-146, 148, 161, 164, 168, 171-173; business, 125; cables, 139; companies, 32, 55; crew, 118, 129; costs of, 34, 123; manager, 106; prospects, 12, 16, 43, 56, 64; prospectus, 22-23, 83, 86; technology, 58; tools, 178
drilling consultants; *see* consultants
drilling contractors; *see* contractors
drilling fluid, 147, 149, 151-152, 161, 169, 172, 174-175
drilling in, *149*
drilling mud, 31, 61-62, 88, 97, 114, 117-119, 129, 138-139, 144-147, 149, 151, 158, 160, *161*, 162, 171-172, 177; additives to, 129; companies, 30, 45; density of, 118-119; viscosity of, 118-119; weight of, 112-114, 118-119
drilling out, *149*
drilling pad, 15, 96; building of, 15
drilling rigs, 5, 15, 32, 44, 58-60, 64, 87, 90-91, 94, 97-99, 106, 113-114, 116-117, 124, 128-129, 143, 148, 169, 172; companies furnishing, 87; moving, 87; sea, 99
drilling slot, *149*
drill pipe, 30-31, 61, 64, 88-89, 114, 119, 129, 138, 140, 142, 145-148, *149*, 150-151, 157, 159, 161, 163, 167-168, 171-174, 176-179
drill pipe joints, 143
drill site; *see* location
drill stem, *149*, 174
drill stem test (D.S.T.), *149*
drill string, 161, 175
drip, *149*
drive bushing, 157
dry gas, *149*
dry hole, 45, 47, 97, 128, 145, *149*
dump bailer, *149*
dump valve, *149*
dynamometer, *150*

earphones, 103
easements, 96-97; *see also* right-of-way
edge water, *150*
Edwards formation, 17
electric logging, *150*
electric pilot, *150*
elevator, *150*
Ellenberger formation, 17
emulsion, *150*
engineering supervision; *see* consultants
engineers, 14-15, 19, 21-23, 58, 74-76, 79, 88-89, 94, 96, 102-106, 108
entrainment, *150*
environmental loss and damages, 33
equipment, 117; cuts in, 111; used, 116-117; wellhead control, 117
equivalent weight, *150*
errors, 97, 111
escalated prices, 30
exchanger, *150*
expansion joint, *150*
expenses, escalation of, 49
exploration, 7-8, 14, 56, 124; vice-presidents of, 12-13, 31
extraction, *150*

fast-buck artists, 12-13
fault, 19, 21-22, 133, *151*
fault line, 151
F.F. flange, *151*
field personnel, 5, 90, 95
field tests, 160
filling the hole, *151*
finger board, *151*
fire wall, *151*
fish, *151*
fishing, *151*
fishing tool, 166, 169, 173
fittings, *151*
flange union, *151*
flare, *151*
flare line, 38
float collar, *151*, 152

—184—

Index

float valve, *152*
flooding, *152*
Florida, 56
Florida Gas Transmission
  Company, 28, 56
flow bean, *152*
flow chart, *152*
flow line, *152*
flowing by heads, *152*
flowing pressure, 49
flowing well, *152*; see also
  producing wells
flow nipple, 152
fluid level, *152*
flush production, *152*
follow-through, 99
footage permits, 172
formation damage, *152*
formation water, *152*
frac fluids, 26, 100, 104
frac job, 22, 32, 102, 104-107, *152*;
  procedure, 100, 102-106, 108;
  service company, 108
frac master, 106
frac tanks, 32-33, 91, 103
fractionator, 145, 148
free gas, *153*
free point indicator, *153*
freshwater sand, see sand
Frio formation, 17
frost box, *153*

gall, *153*
gallons per thousand (G.P.M.),
  154
gas cap, *153*
gas drive, *153*
gas flow, 26
gas lift, *153*
gas lock, *153*
gas metering charts, 114
gas-oil ratio (G.O.R.), *153*
gas liquid ratio, *153*
gas pipeline; see pipeline
gas valve, *153*
gate valve, *153*, 178
gathering lines, *153*

gauge, *154*
gauge glass, *154*
gauge hatch, *176*
gauge pressure, *154*
gauger, *154*
generators, 116
geological formations, 17-18,
  21-22; structure of, 19
geologists, 17, 19, 22-23, 44-45,
  88-89
gin-pole, *154*
globe valve, *154*, 178
glycol, *154*
go-devil, *154*
Gonzalez County, Texas, 23
good ole boys, 13, 15, 128
goose neck, *154*
grade, *154*
gravel pack, *155*
grind out a sample, *155*
guide shoe, *155*
Gulf of Mexico, 56
gun-barrel, *155*
guy, *155*

Haliburton, 108
hauling oil, 93
hauling tickets, 93
head, of liquid, *155*
header, *155*
heat crystallization, 117
heirs; see mineral leases
hold down nipple, *155*
holiday, *155*
hot spot, *155*
Houston, Texas, 26-27, 33, 74-76
Hughes, B.J., 108
hunting, *156*
Hunt Oil Company, 129
hydrafrac, *156*, 173-174
hydrate, *156*
hydrogen sulfide, *156*
hydrostatic presssure, *156*

impervious, *156*
impression block, *156*
incompetence, 96-97, 99-100, 104,

Index

106-108, 111; in well stimulation, 100
independent oil companies; *see* oil companies
independent operators, 33-34
inflation, 123
injection pressure, 100, 103
injectivity test, *156*
insurance, 112, 116, 118; adjuster, 118; representative, 120
integrator, chart, *156*
interests, in oil venture, 8, 45, 86, 97, 124; to casing point, 9, 127; to completion point, 9, 127; to logging point, 9, 127; to sales point, 9, 127
intermediate string, *157*
Internal Revenue Code, 123
investment points, 127
invoices, 74, 76, 126, 131
irregularities, 132
itemized statement of expenses, 132
jack-screw, *157*
jack-shaft, *157*
Jackson County, Texas, 113, 128
James Lime formation, 17
jars, *157*
jerk lines, 143
jet perforating, *157*
job completion; *see* completion
joints, *157*
journalists; *see* public relations
"J" slot, *157*
junk mills, 161

kelly, *157*, 167, 169, 175
kelly bushing, *157*
kelly cock, *157*
key seat, *157*
kickback, 13, 17, 86, 91
killing a well, *158*
Kurten, Texas, 86

land department; *see* oil companies
land developing companies, 99
landholders, damage to, 5
land records, 31
lawsuits, 5, 97,112, 118, 134
land casing, *158*
latch on, *158*
laterals, *158*
lead acetate test, *158*
lease; *see* mineral lease
Lease Automatic Custody Transfer Unit (L.A.C.T.), *158*
leaseholders, 4, 44-45, 98-99
leasehound, 8, 17
legal fees, 6-7
lending companies, 30
lessee, 3, *158*
lessor, 3, 96, 132, *158*; royalty to, 8
liens, 73-74, 134
lifting cost, *158*
limestone, 102
line out, *158*
liner, *158*
Liquified Petroleum Gas (LP gas), *158*
lis pendus, 73
load factor, *158*
load the hole, *159*
location, 44, 47, 70, 94, 96-97, 103, 118, 120, 159; building roads to, 86-87; cleaning up, 32, 94, 97; construction of, 15, 32, 86, 94, 96-97, 108-109; rigs, 63, 106
log, *159*, see also well log
logging; *see* well log
logging graph, 71
logging point, 124, 127
log library, 56
loop, *159*
loopholes, 131-132
lost circulation, *159*
lost returns, *159*
lubricator, *159*

make tank, *159*
make-up, *159*
M-alkalinity test, *159*
manifold, *159*
Marble Falls formation, 17

—*186*—

Index

marginal well, *160*
marine-life, protection of, 97
marketing agents, 8, 125
marketing costs, 6-7
marsh funnel, *160*
mast, *160*
master bushing, 169, 172
material balance, *160*
Maximum Efficiency Rate (M.E.R.), *160*
meter run, *160*
migration, *161*
mill, 106, *161*, 170
Millican Oil Company, 90
mineral lease, 3, 12-13, 22, 31-33, 36, 38, 44-47, 56, 58-60, 62, 74, 86, 91-92, 96-99, 116, 118, 127-129, 131-133, 135; costs of, 7; division of, 7; drilling contract method of, 3, 44; escalation of price on, 11; excessive profits on, 6; heirs in, 4; hidden costs in, 6; lessee, 3; lessor, 3; per-acre bonus for, 3, 6, 45; problem on, 5; rental fees, 3; terms of, 3
mineral rights, 4, 44, 132
mistakes, 96-97, 111
mist extractor, *161*
monoethanolamine (MEA), *160*
money disbursements, division orders for, 14
Mobil Oil Company, 129
Mohr test, 144
monitoring van, 103
mouse hole, *161*
MSCFD, *161*
mud, *161*; *see also* drilling mud
mud cake, *161*, 170
mud hose, 169
mud logging, 125, *161*; *see also* well logs
mud pits, 15, 32-33, 96-97, 105, 119; digging of, 15, 33; closing of, 32
mud pump, *161*, 169
mud screen, *162*

mud weight, *162*; *see also* drilling mud

natural flow, *162*
natural gas, 19, 38, 53-54, 91, 137, 149, 156, 158, *162*, 179; water content of, 19
needle valve, 178
Net Operating Loss Deductions (N.O.L.), 123
New York, 36-37
nipple, *162*
normal bottomhole pressure, *162*; *see also* bottomhole pressure
nozzle, *162*
nutating disc, *162*

offset well, *162*
oil-bearing zones; *see* production zones
oil column, *162*
oil companies, 4-5, 43, 47, 68, 86, 90, 97-98, 106, 112, 114, 116, 118, 126, 132, 134-135; boards of, 30-31, 88, 125; chairmen of boards of, 30, 125; division order department of, 31; employees of, 28; engineers of, 108; executives of, 27, 29, 75; expenditures of, 82; furnishing rigs, 87; independent, 6, 25, 56, 74-76, 125; land departments of, 12-14, 31, 99; large corporations, 30; major, 25, 46, 56-58, 63, 74-77, 135; management of, 8; marketing people of, 6, 32; offices of, 27, 29; order records of, 31; overcharging for pipe, 88; presidents of, 5, 90, 120; private corporations, 25, 35; production departments of, 31; small, 25-27, 30, 32-35, 56, 74, 113, 125-126, 135; stock-option programs of, 27-28; vice presidents of, 90, 104
oilfield supply houses, 135

—187—

Index

oilfield workers, 5
oil reserve, estimated, *162*
oil saver, *162*
oil shortages, 93
oil storage, 23, 72, 93
oil string, *162*
oil squeeze, *163*
oil well completion; *see* completion
oil well drilling, *see* drilling
Oklahoma, 64
open ended, *163*
open flow test, *163*
open hole, *163*
open hole logs; *see* well logs
operating capital, 126
operating company, 32
operating expenses, 10, 12, 49, 51, 167; prorated share of, 12
option on ventures, 34, 92
orifice meter, 147, 160, *163*
outage, *163*
overburden pressure, *163*
overcharges, 32, 111
overpayments, 16, 86; investment of, 15
overriding royalty interest, 6-9, 11-14, 31, 44-45, 55, 67, 133; assignment of, 13
overshot, *163*

packer, 102-105, *163*, 171-172, 174-176
P-alkalinity test, *163*
paraffin, *163*
Parks and Wildlife Agencies, 117
pass, *164*
payout, *164*
percentages and disbursements, 9
perforate, *164*
perforating gun, 75, 77, 79, 157, 174,
perforating of wells; *see* well-perforation
perforating truck, 75, 79
permeable, *164*

pipe, 30-31, 37-38, 59-60, 77, 88, 100, 105-106, 111-112, 116-117, 120, 128, 138, 141, 144, 149, 153, 155, 160, 162-163, 169-170, 172-178; overcharging for, 88; production, 59; *see also* drill pipe, surface pipe
pipe coupling, 145
pipe fittings, 144, 164
pipe header, 159
pipeline, 24, 50, 54-56, 91-92, 114, 131, 133, 138, 150, 152-154, 156, 158, 163, 169, 176; crude oil, 55
pipeline carrier, 91
pipeline companies, 54, 91, 114; liens on, 73-74
pipeline connection, *164*
pipeline oil, *164*
pipe rack, 178
pipe suppliers, 88, 135
pipe welding, 164
platform, *164*
plug, *164*
plug back, *164*, 179
plugging of holes and wells, 32, 47, 53-55, 59-62, 72, 77, 97, 128-129
polished rod, *164*, 174-175
polished rod clamp, *164*
pop valve, *165*, 170
pore space, *165*
porosity, *165*
potential test, *165*
pounding fluid, *165*
power tongs, 176
present worth, *165*
pressure base, *165*
pressure build-up, 106
pressure gauge, 104, *165*
pressure maintenance, *165*
primacord, 174
primary production, *165*
producing wells, 5, 8, 12, 17, 22-23, 35-36, 43-44, 46-47, 54, 58-59, 62-63, 67-70, 72-74, 93, 104-105, 116, 118, 123-124, 126, 131, 133-134, 144, 164

—188—

Index

production, *166*; facilities, 23, 38, 49, 72; rate, 51, 76, 102; records, 31; reports, 26, 133
production casing, *see* casing
production curves, *166*
production department; *see* oil companies
production engineer; *see* engineers
production pipe; *see* pipe
production tubing; *see* tubing
production zones, 17, 20, 22, 45-49, 51, 57-58, 60, 62, 75-77, 88-89, 100-102, 112-113, 129; calculated depth of, 22; chalk, 21; clay, 21; limestone, 21; permeability of, 19; porosity of, 19; sandy, 21; thickness of, 19
productivity index (P.I.), 166
profitability index, *166*
propane, 38, 158
propane tanks, 38
proppant, *166*
prorate, *166*
public relations, 25-26, 38, 125
pulling costs, *166*
pulling line, *166*
pulling tool, *166*, 167
pulling unit, *178*
pump, 116, 161, 166, 169
pumper, *167*
pumping operation, 105
pumping pressure, 100, 102, 105
pumping procedure, 103
pumping rate, 100
pumping unit, 49, 104, 106
pump lines, 106
pump trucks, 103

racking pipe, *167*
radios, 103
railroad, 118
Railroad Commission of Texas, 23, 75-76
rams, 119, 139-140
rat hole, *167*
reaming, *167*

reaming mills, 161
recompletion, *167*, 179
recorders, 103
recovery period, 133
re-entry, 52, 55-65, 129; dangers of, 60
refineries, 55
regulator, *167*
repressuring, *167*
re-run, *167*
research and development, 53, 103, 109
reserves, 9, 48, 51, 54, 118, 133
reservoir, *167*
reservoir energy, *168*
resident lessors, 5
resources firm, 31
responsibilities, 111, 113, 116
retention, *168*
retrievable, *168*
return on investment (R.O.I.), *167*
revenue, 49
revenue circulation, *168*
R.F. flange, *168*
rig; *see* drilling rig
rig company, records of, 109
rig crew, 105, 112
rigging up, *168*
right-of-way, 4, 96-97; *see also* easements
rig site; *see* location
rig supervision; *see* consultants
roads, access, 96-98; construction of, 15, 86-87, 94, 96-97
rocking the well, *168*
rock pressure, *168*
rod, 104, 106, 138, 164, 166, 179
rod guide, *169*
rotary bushing, *169*
rotary drilling, *169*
rotary hook, 175
rotary hose, 175
rotary house, *169*
rotary slips, 172
rotary table, 157, 169
rotary tongs, 176
roughneck, 62-63, 148, *169*

—189—

Index

round trip, *169*
roustabout, *169*
royalties; *see* overriding royalty interest, working interest
royalty pool, 8, 14
ring type joint (R.T.J.), *169*
run, *169*
run-tickets, 70, 133; audit of, 93

saddle, *169*
safety head, *169*
safety joint, *169*
safety latch, *170*
safety valve, *170*
sales commission, 39, 67-69
sales line, 9, 24, 36, 43-44, 50, 54, 56, 127, 131, 133-134; *see also* pipeline
sales pitch, 16, 38
sales point, 16, 54, 133
sales volume, 50
salt cavity, *see* cavity
salt water, 50, 72, 116, 137
salt-water tanks, 72
sand, 51, 102, 104-105, 113-117, 120, 138, 147, 149, 156, 170; freshwater, 59, 112; mixture, 105
sandblasting, 113
sand formations, 170
sanding up, *170*
sand pump, 138, *170*
sandstone, 102
sand tanks, 103
scale, *170*
scams, 35, 67, 70; *see also* schemes, swindles
schemes, 89, 92
scratcher, *170*
screen, *170*
scrubber, *170*
section milling, *170*
seismic surveys, 19
separator, *170*; oil and water, 23
series, *170*
service companies, 103, 105; personnel of, 108; records of, 109

settled production, *170*
shake-out, 93
shale, 51, 56
shale shaker, 34, 162
shallow wells, 49, 55
shell pump, 170
shooting, 179
shot feeder, *171*
show, *171*
shrinkage, *171*
shut-in tubing pressure, *171*
shut-off, *171*
Siboney Petroleum Company, 129
side-tracking, 61, 161, *171*, 179
sidewall coring, *171*
single, *171*
skid, *171*
skimming pit, *171*
slacking off, *171*
slips, *171*
slip tube, *172*
sloughing, *172*
sludger, 170
slug, *172*
slug the pipe, *172*
slurry, *172*
slush pit, *172*
slush pond, 172
small oil companies; *see* oil companies
solution gas-oil ratio, *172*
sonde, *172*
sour gas, *172*
sour liquid, *172*
sour oil, *173*
South Texas Oil and Gas Company, 129
spacing, *173*
spear, *173*
spider, *173*
spinning line, *173*
spool, *173*
spot oil, *173*
spotting, *173*
spudding in, *173*
squeeze cementing, 179
squeeze job, *173*

—*190*—

Index

stabbing, *173*
stabbing board, *173*
stabilizer, *174*
stand, *174*
standard conditions, *174*
standing valve, *174*, 179
standpipe, 169, *174*
statement of expenses and sales, 35
static pressure, *174*
stiff leg, *174*
stimulation procedures; *see* well stimulation
stock, shares of, 28, 35-36, 38-39
stockbrokers, 26-27, 39
stock exchanges, 26-28, 30-31, 35-36
stockholders, 30-31, 36; abuses of money, 30-31
stockholding companies, 27, 31-33; organizational structure of, 31
stock option plans, employee, 27-29; minimal payment on, 28; monthly deductions for, 28
stop-cock, 174
stop-cocking, *174*
stopping point, 64
storage facilities, 70, 139, 172; *see also* tanks
straddle packer, *174*
strap, *174*
string shot, *174*
stripper, *174*
stroke, *174*
stuffing box, *175*
sub, *175*
substructure, *175*
sucker rod, 49, 139, 150, 157, 162-163, 169, *174-175*
sucker rod hanger, *175*
sucker rod jack, *175*
sucker rod pump, 165, 178
sucker rod string, 164
supply and demand, 53
surface land rights, 4, 98, 132
surface pipe, 60, 111-112, 114, 116, 128-129, *175*
survey, 99, 148
swabbing, *175*
swab lines, 162
sweet crude, *175*
swindles, 39
swivel, *175*

tail pipe, *176*
tally, *176*
tank battery, *176*
tank bottoms, *176*
tank cars, 172
tank gauge, *176*
tanks, production, 38, 134, 145, 170, 176; propane, 38
tap, *176*
taxes, 10, 12, 49, 68, 123-124, 126-127, 167
tax incentives, 123-124, 127, 134
technical supervision; *see* consultants
temperature, 61
temperature survey, *176*
Tennessee Gas Transmissions, 54, 56
testing reports, 56
test results, 25
Texas, 35, 37-38, 49, 51, 54, 74, 76, 87, 111, 117-118, 142
Texas Gulf Coast, 21
thief, *176*
thief hatch, *176*
throttle, *176*
tie-down, *176*
tongs, 138, 173, *176*
tool joint, *177*
toolpushers, 44, 88, 90, 105, 113-114
total depth (T.D.), 43, 127-128, *177*
traveling block, 139-140, 175, *177*
Travis Peak formation, 17
treating fluids, 152
trespassing, 96, 98
tribble, *176*
truck hauling, 23, 55, 93; independent, 15

—*191*—

Index

trust officers, of banks, 68-69
tubing, 49, 100, 102-105, 138, 140, 142-143, 147-150, 159, 161, 163, 166-167, 169, 171, 173-174, 176-177, 179
tubing anchor, *177*
tubing catcher, *177*
tubing contracts, 88
tubing hanger, 177
tubing head, 144, 177, 178
tubing string, 155, 177
twist off, *177*

underream, *177*
unload the hole, *177*
Utah, 39
utilities, 118

valve, 37, 91-92, 117, 138, 141-142, 146, 153-154, 157, 159, 165, 170, 174, 176, *177*
V-door, *178*
Victoria, Texas, 115
viscosity, *178*

waiting on cement (W.O.C.), *178*
walking beam, 139, *178*
Wall Street Journal, 26
washover, *178*
wash pipe, 104, 178
water column, *162*
Water Control Board, 111-113
water-disposal cost, 50
water drive, 51, *178*
water flooding, 49
water hardness, *155*
water knockout, *178*
water pits, digging of, 15
water saturation, 61
watershed area, 97
water supply, 111
water-table, *178*
water wells, 96, 148
welding, 114, 116
well completion; *see* completion
well control procedures, 120
well depletion ; *see* depletion

well drilling; *see* drilling
wellhead, 30-31, 37-39, 88, 102-103, 106, 114-115, 117-118, 159, 171, 175, 177 ; business, 125; cemented, 36, 38; contracts, 88; control equipment, 117, 120; dummy, 39; pressure on, 37, 168
wellhead assembly, *178*
wellhead gauge, 37
well-logging companies, 45
well-log, 19, 21, 44-45, 58-60, 62, 64, 112, 124, 127-129; old, 56, 58-60, 62, 64, 74-75, 77, 79, 89
well-perforation, 48, 75, 77, 79, 100, 102
well-plugging procedures; *see* plugging of holes and wells
well production; *see* producing wells
well-production reports; *see* production
well servicing unit, *178*
well-site; *see* location
well-stimulation, 107-109; incompetence in, 100; procedures, 57, 100, 102-103
well surveying, *179*
well tests, 56
well tester, 25
Western, 108
wet gas, *179*
wet string, *179*
whipstock, *179*
Wilcox formation, 17
wildcat, *179*
wildcatters, 25, 55-57
wild well, *179*
wild-well fighters, 117-118, 120
working barrel, *179*
working capital, 8
working interest, in wells, 6-8, 10, 12, 15-16, 43, 45, 47, 49, 99, 124-125
workover, 72, 131-132, *179*; procedure, 132,; costs of, 131
workover rig, 49, 59, 72, 90, 109

—192—